MW00811308

Math Learning Strategies

Math Learning Strategies

How Parents and Teachers Can Help Kids Excel in Math

Teruni Lamberg

ROWMAN & LITTLEFIELD
Lanham • Boulder • New York • London

Published by Rowman & Littlefield
An imprint of The Rowman & Littlefield Publishing Group, Inc.
4501 Forbes Boulevard, Suite 200, Lanham, Maryland 20706
www.rowman.com

86-90 Paul Street, London EC2A 4NE

British Library Cataloguing in Publication Information Available

Library of Congress Cataloging-in-Publication Data Available

ISBN 9781475867237 (cloth) | ISBN 9781475867244 (ebook)

Dedicated to my parents—Kenneth and Shireen de Silva—
and to my husband.

Contents

Acknowledgments

The goal of this book is to make a difference in the lives of many children so that they can reach their potential in learning math. Writing this book was a journey of love that involved the support of many individuals who inspired me and shaped my thinking. I am thankful to all the participants in my research study who shared their experiences and insights on how to support students in learning math.

The following individuals gave me valuable feedback on my ideas and shaped my thinking through the writing process: Seth Murphy, Ethan Ris, Kevin Kalra, Diana Moss, Gideon Caplovitz, Craig Wall, Betsy Hawkins, Linda Taylor, Tracy Gruber, Riley Gruber, Jeremy Jimenez, Edward Keppelman, Alysia Goyer, Diana Moss, Ethan Ris, Tom Quint, Tin-Yau Tam, Rachel Hawkins Sipe, Peggy Lakey, Alysia Goyer, Ana Bettencourt-Dias, Sarah Cummings, Lissa Kellogg, and Tawnya Michie Kumarakulasingham. I am thankful for their time and expertise.

I am thankful to all my students, who inspire my work every day. Thank you to Betsy Hawkins and Elaine Pohle, who helped me edit my manuscript. I appreciate them taking the time to give me constructive feedback on my writing. A special thank you to Tom Koerner, Carlie Wall, and Tricia Currie-Knight at Rowman & Littlefield.

My parents, Kenneth and Shireen de Silva, shaped my journey as a teacher, scholar, and author. I am so grateful for their love and support. My husband and son always kept a smile on my face with their sense of humor. I am thankful for their support.

Introduction

Jamie was excited to attend the parent-teacher conference. Wesley was doing great in everything, but the teacher pointed out that he struggled in math. He was barely passing the class and scored low on the state test. Jamie went home anxious, not knowing what she could do to help Wesley. As a parent, she felt hopeless. She wanted Wesley to do well. The worksheets he brought home did not make sense to her. The teacher said they were doing the "new math" in the Common Core Mathematics Standards and told her that she needed to teach him at home. When Jamie tried to make sense of the worksheets that Wesley brought home, she did not know how to help. The math did not seem familiar to how she learned math at school.

Night after night, she tried to show him the "steps" to solve the math problems. Wesley seemed unmotivated, and the battle to complete the homework ensued. After a long day of working, cooking dinner, cleaning the house, and taking care of the baby, she was exhausted and did not look forward to the battle of getting Wesley to do his homework. She would have rather spent her time having fun with him instead.

Wesley was tired and worn out from the day and did not look forward to doing math either. He spent all evening trying to complete the worksheets at home. Wesley was unmotivated to complete them and exhausted by bedtime. He did not have time to play, rest, and recharge before he went to school the next day. Wesley would often say that he completed the homework at school. However, Jamie discovered from his teacher that he did not turn in all his homework. Usually, the worksheets seemed to disappear on the way home from school. Wesley just did not care.

When Wesley was born, Jamie dreamed of wonderful things for her son. She envisioned a future where he could attend college or get a technical job. Jamie knew her child's future rested on his ability to do math because his ability to pursue higher education and STEM (science, technology, engineering, and math) jobs required doing well in math. When she looked at the state test score data, she was alarmed to know that her child was not the only child struggling in math. Other parents shared the same stories with her. Her state's proficiency rates in math were low for elementary school and even lower in middle and high school. This data scared her. She wanted to do something about it to help her child love learning.

The teacher also cared about Wesley and wanted him to learn. However, she did not know what to tell his parents or how to support him. The teacher wondered how to work with Jamie to help Wesley. Over the years, many parents shared that they were not comfortable or good at math themselves. Therefore, she needed to think about how to work with the parents to help their children effectively. Jamie wondered what she could do to spark Wesley's math brain to develop a love of learning math and succeed.

Several parents have shared similar stories and concerns with me over the years. Therefore, this book provides parents and teachers concrete ideas to help students learn math. Specifically, parents and teachers will gain insights on how to help students learn math by building collaborative relationships to support their students. Parents will know what to consider when talking to teachers and how to advocate for their children. Teachers will learn about the kinds of support parents find helpful at home. In addition, teachers will learn strategies on *what* and *how* to communicate with parents.

This book aims to help students become successful in math to reach their potential. Unfortunately, low performance in math prevents many students from pursuing careers requiring higher-level math. Therefore, this book outlines problem-solving strategies students can use to make sense of math problems and understand the math. In addition, it outlines the key mathematical concepts that students need to understand in grades K–5 based on the Common Core Mathematics Standards that many states have adopted.

Sometimes the path to learning math may not be linear, and many obstacles can happen along the way. The reality is that each child is unique and has their own set of gifts and talents. Each student must figure out what works for them. Therefore, my companion book *Sparking the Math Brain: Insights on What Motivates Students to Learn—Creating Conditions for Learning* explores topics of motivation and perseverance to keep a child on track.

We can think about success beyond academic success. Living a fulfilled life involves doing what you love, living a good life and being happy, and contributing to society. Academic success and being a high achiever may not contribute to these ideals. Therefore, each family and student must discover and decide what they value and what works for them. Many individuals who did not achieve academically make valuable contributions to society every day.

I am a former elementary teacher, a college professor in mathematics education, and a parent. I met many mothers with the same anxiety as Jamie over the years. My 21-year journey to help children learn math has involved working with hundreds of teachers across the country. Many teachers lament that they would love to see more parent involvement. Yet, many teachers struggle to keep up with the demands of teaching and find time to communicate and work with parents.

This book contains original research. Twenty-one individuals who pursued a master's degree or doctorate or attended an Ivy League school were interviewed. Seven individuals interviewed were mathematicians, and five were STEM graduates (one engineer, one doctor, and three scientists). Four individuals interviewed pursued non-STEM degrees (humanities, geography).

Fifteen successful adults graduated from Ivy League schools, such as Stanford, Harvard, Yale, Princeton, or Columbia, or other high-research institutions. These 15 individuals successfully met the math requirements to enter college and pursue higher education in STEM or other disciplines. They learned how to navigate school and do very well in math. The adults who completed a master's degree or doctorate in STEM disciplines provided insights into what sparked their love of math and perseverance to become mathematicians or scientists.

The interviewed individuals represent different geographic areas (United States, Europe, and Asia), including rural, urban, and suburban regions and diverse socioeconomic backgrounds. The geographic areas of the United States represented include the West, Midwest, Northeast, Southeast, and Southwest. The individuals ranged in age from mid-twenties to sixties. They were at various stages of their adult careers.

In addition, five parents of adults who pursued a master's or doctoral degree in science, technology, engineering, or mathematics were interviewed. These parents were from the United States and represented urban, rural, and suburban backgrounds. The goal was to gain insights into a parent's perspective on what they explicitly did to support their children. The parent's perspective provides insight into parents' intentional actions to help their children in math.

The hour-long interviews were transcribed and analyzed for themes using the Constant Comparative method (Strauss and Corbin, 1998). The interviews focused on the questions listed in the appendixes. A snowballing methodology was used to encourage individuals to elaborate on their responses. Each chapter contains themes that emerged from the data. Pseudonyms are used throughout the book when sharing excerpts from interviews.

The findings reported in this book focus on learning strategies and environments that support a high level of math learning. The fictitious characters of

Jamie and Wesley represent scenarios many parents wonder about as they try to help their children. As a result of reading this book, parents and teachers will gain insights and strategies to help their children learn math more effectively by focusing on parent and teacher partnerships, learning strategies, and how to create conditions to support math learning.

BIBLIOGRAPHY

Strauss, A. L., & Corbin, J. (1998). *Basics of qualitative research*. Sage.

Part I

CREATING CONDITIONS FOR HELPING STUDENTS TO LEARN MATH

Chapter One

What Does It Mean to Learn Math?

Learning happens when you connect new information to what you already know. When you can explain your thinking to others and use that information in a new problem situation, learning has taken place!

* Doing mathematics is like solving a puzzle or playing a game!

> Wesley stared intently at the chessboard to plan his next move, in order to beat his opponent. This situation was a stressful moment as he pondered what to do next. There are several possibilities for action. He imagined each scenario in his head and evaluated the pros and cons. If he moved the rook, he could capture the knight. However, that move would leave his queen vulnerable. Suddenly it became clear, and he saw what he had to do! He moved his queen and yelled out, "Checkmate!" with a feeling of joy!

Solving a problem that has a level of challenge can be stressful. Individuals experience joy when they can solve a problem. The mind analyzes the problem, seeks patterns, and produces a solution. Logical and critical thinking is involved when choosing the best solution based on the situation. The process of figuring things out and noticing patterns results in a feeling of accomplishment and satisfaction despite the initial struggle. Individuals who love math experience math as solving puzzles.

Mathematics is not much different than playing chess. Research has shown that individuals who do well in mathematics experience joy while doing mathematics and treat mathematics as a pleasurable experience like solving a puzzle and playing logic games (Rizzolatti et al., 2007). Mathematicians typically break larger problems into smaller parts and tackle each piece simultaneously. Solving problems with the right level of challenge can be a joyful experience, similar to solving a puzzle. When mathematicians experience doing math, like a solving puzzle, it engages the brain. The individual must think about the problem, look for patterns, and understand the information provided. When people experience joy, they are motivated to learn.

Understanding how learning happens and what it looks like is fundamental for helping children learn math. Think about a time when you learned something. What did you experience? How did you know that learning took place? The National Research Council (2000) wrote a book called *How People Learn: Brain, Mind, Experience and School (Expanded Edition)*. In this book, they describe how learning and *transfer* happen. They point out that learning occurs when there is *understanding*.

It is not important that the student memorize that 25 + 25 is 50. Rather, a student must *understand* why adding two 25s together makes up 50. There are several ways a student can think about how to add the numbers together. The student can use blocks to count out 25 blocks by using two tens place value blocks and another five individual blocks. Another method might involve doubling the 20 and adding the two fives together. The student can also use the standard algorithm of borrowing and carrying, as illustrated in figure 1.1 below.

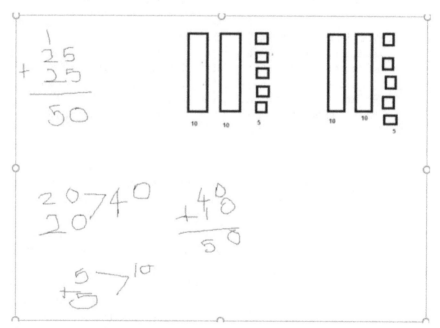

Figure 1.1.   Multiple Ways to Add 25 + 25
Teruni Lamberg

- Deep learning happens when students understand the underlying concepts of what is taking place.

Deep learning happens when the student understands the underlying concepts of what is taking place. For example, when the student can see how different strategies connect, the student develops a deep understanding. To understand how to add these two-digit numbers, the student must understand how to count and model the numbers using manipulatives. In addition, they should be able to compose and decompose numbers (break numbers apart to make friendly numbers) and place value (ten's digits and one's digit), and to regroup and represent the problem using standard math notation. (See figure 1.2.)

A concept map represents the relationship between concepts and how they are linked (Chiou, 2008). Concept mapping is helpful for students to see math as a network of interrelated concepts. Concept maps are beneficial in math because they help facilitate meaningful learning by assisting students in organizing their ideas, which helps them remember (Brinkmann, 2003).

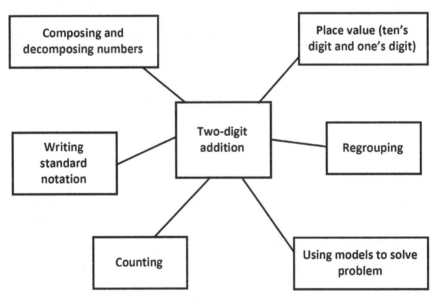

Figure 1.2. Concept Map for Two-Digit Addition with Regrouping
Teruni Lamberg

- If learning takes place, students can use what they know to solve other problems.

When we think about learning, we need to think about the *transfer*. In other words, it does not matter if a child figured out that 25 + 25 is 50. Memorizing the answer without understanding why 25 + 25 is 50 is meaningless. A child may remember 50 in the short term, but forget it later because the answer did not make sense. If the student understood the underlying concepts of adding two-digit numbers together, then when given a similar problem, such as 25 + 28, the student should be able to draw from their prior knowledge. For example, the student might see that 28 is three more than 25. So, the answer is 53.

- The key to learning is the ability to see connections and patterns. When something makes sense, it is easier to remember.

According to the National Research Council (2000), experts notice patterns because they understand content deeply and meaningfully. For example, they point out that expert chess players can see patterns on a game board in a meaningful way, as opposed to novice chess players, who see pieces on the board. Therefore, we must help our children develop the ability to see patterns.

TURNING ON THE MATH BRAIN

- Learning math is more than memorizing a bunch of facts and formulas; it is about developing understanding.

Even though Jamie was frustrated that her son Wesley did not do well at math, she knew in her heart that Wesley loved playing Monopoly for hours and was able to count out change and add money. He knew how to add the Monopoly money effortlessly. She knew that Wesley's test scores did not reflect his real ability. She wondered why he was failing math, especially when she observed that he loved playing games involving math. Why did Wesley love playing games but not doing schoolwork? It was because Wesley did not see school math as a pleasant experience that had anything to do with real life or the games he played. Instead, he dreaded it because he was not good at memorizing formulas and hated doing meaningless math homework.

What children learn in natural settings may not translate into school math. Carraher et al. (1985) conducted a study by observing street children in Brazil selling items and engaging in commercial transactions. During this process, the children could do sophisticated math for the transactions in an authentic way. However, when these children were given similar problems in school, they did not perform well.

Therefore, the study concluded, "performance on mathematical problems embedded in real-life contexts was superior to that on school-type word problems and context-free computational problems involving the same numbers and operations" (p. 25). These children used different approaches in their environments. They did not see a connection between the informal setting and the formal school. Wesley's mother had to figure out ways to help him see connections between the games he played and school math. Wesley needed to see that doing math can be fun and meaningful.

MATHEMATICIANS' PERSPECTIVES ON THE BEAUTY OF SOLVING A MATH PROBLEM

- Doing mathematics involves breaking things apart, solving parts of the problem, and putting things together to get the answer.

Mathematics can be a joyful experience when treated as a puzzle or something interesting to solve. John, a mathematician, explained that he experienced *joy* when solving problems. He spent his time looking at the big problem, breaking it down into parts, and focusing on one part at a time. During that time, John looked for patterns and persisted in solving the problem. Sometimes he would set it aside and revisit it with a fresh, new perspective. He explained, "I guess I really got excited about figuring things out and putting order into a situation that appeared chaotic—figuring out how all the pieces fit together."

Solving math involves understanding math and having strategies to solve problems, the ability to monitor oneself, impact what mathematicians do (Schoenfeld, 2011). Emotions impact thinking and learning, too (DeBellis & Goldin, 2006).

- Logical thinking and problem solving reveals the beauty of mathematics!

The process of engaging in logical thinking is beautiful. Ying, another mathematician, shared that he started to see the beauty of mathematics in high school when he was learning calculus. He learned to do proofs through the epsilon delta, a classical analysis approach. He was fascinated by logical thinking and the beauty of mathematics.

> You may not believe that, but the teacher taught calculus using the epsilon delta approach. So, they used the so-called classical analysis approach called epsilon delta to teach calculus, which involves proofs, so I was fascinated by it.

I was fascinated by the beauty of mathematics and the argument. The way he handled mathematics that way—the logic, or the vigorous proof. And I picked that up, not immediately, but I picked up pretty quickly.

• Mathematics has a structure where each step builds on the previous steps.

There is a concrete nature to math shared by mathematicians. "In other words, there is a beginning, middle, and end when solving a math problem, and each step builds on each other to get an answer." Ying pointed out that solving math problems requires understanding that the structure builds on itself. Learning math is cumulative and builds on prior knowledge across grade levels. He pointed out that this is not the same in other disciplines such as reading and writing; the order of learning things does not matter as much.

• Mathematical solutions are aesthetically beautiful!

A mathematical equation written out elegantly is visually beautiful. Ying explained, "The beauty of math is the mathematical relationships; the actual mathematics itself is beautiful—the way things fit together and this logical progression to things. There is an aesthetic quality to math where it looks cool." He shared that in history there are stories of how the beauty of math was valued. "There are these stories of an artist who came to the emperor's court and drew a perfect circle for the emperor and was rewarded with treasures." Students need to be inspired to see the beauty of mathematics.

MATH IS FUN AND COOL!

The professor confidently walks into the classroom with his big hair and cool outfit. He loves to use multicolored chalk. He walks over to the board and pulls down one board so that he can write on it. He picks up the chalk, and his hand glides elegantly across the board as he makes a beautiful perfect sine wave. His body moves dramatically with the grace and flair of a martial artist, as the class silently watches him. They are mesmerized, as he continues to slowly and methodically write a mathematical equation with impeccable handwriting and precision.

The former high school student's interest was piqued as he suddenly became intrigued by the coolness of math. He began anticipating what the professor might do next and worked a little ahead of him. When he got the answer right, he made a checkmark next to it and said to himself, "Nailed it!" while doing a happy dance in his head!

This high school student, who later got his undergraduate and master's degrees in math and continued to earn his doctorate in psychology, suddenly started to enjoy math because it made sense. His interest was captivated by the showmanship and beauty of mathematics. Before that, even though he knew he was good at math, he had checked out of high school as a student. He barely turned in his assignments and did not enjoy math class. Most of all, he did not care.

He blamed the teacher for his getting zeros and later reflected that he did not do his homework and was distracted by the social scene. This class changed his perception and sparked his love of math. He was inspired to take more classes, which resulted in getting his undergraduate and master's degrees in math and then pursuing a doctorate in psychology. In this class, he shared, "I really got to see what math is supposed to look like." Some students are naturally good at math and start loving it early. Other students, given the opportunity, can develop a love of math and become good at it. The school and home environment have the potential to nurture and build on these abilities.

MATHEMATICALLY GIFTED STUDENTS

Mathematically gifted students describe loving math and are naturally good at math at an incredibly young age. These children are so fascinated by patterns that they learn math quickly, and it comes easy to them. Identifying gifted children early and supporting their mathematical abilities sets them up for success. Parents can request schools test their children for giftedness if they find that their children are not being challenged and supported in a regular classroom setting. Gifted programs help children who require a higher level of challenge. Characteristics of gifted and talented mathematics students are a fast-learning pace, exceptional memory, extended concentration span, and curiosity (Singer et al., 2016). Jose, a Stanford graduate, shared:

> I was really, really, really good at math. It came naturally to me; I always won the math league contests. I remember, though, being in second grade, they gave me this big math workbook for the year . . . and I actually finished the work within November.

Mathematically gifted students may also show that they can engage in sophisticated mathematical reasoning and are exceptionally good at problem-solving. In addition, they have good spatial reasoning abilities (Sowell et al., 1990).

Sometimes, mathematically gifted students with strong spatial ability might be overlooked for gifted programs because their test scores may not reflect their ability (Mann, 2004). Teachers and parents should pay attention to students' needs and provide support if gifted services are needed.

SUMMARY: WHAT DOES IT MEAN TO LEARN MATH?

See table 1.1 for lists of goals for parents and teachers.

Turning On the Math Brain
• Students understand the concepts of what they are learning.
• Students understand what they are doing and why the strategies work.
• There is a logical progression when solving math problems. Each part builds on the previous part.
• Treat math learning like solving a puzzle.
• Break the larger problem apart, tackle parts one at a time, and put it all together.
• See the beauty of mathematics: logical thinking and visual aesthetics.

Table 1.1.   What Does It Mean to Learn Math?

Goals for Parents	Goals for Teachers
• Understand what is involved in learning math.	• Help students see the beauty and joy in solving math problems.
• Recognize that doing mathematics is like solving a puzzle or playing a game!	• Teach with a focus on conceptual understanding.
• Help your child discover the joy of doing math. If a child exhibits mathematical giftedness, the child should be tested and placed in an appropriate learning environment.	• Teach through a problem-solving approach.
	• Build on students' prior knowledge.
	• Help them see patterns and connections to what they already know.
	• Get students excited about math by communicating a passion for it.

BIBLIOGRAPHY

Brinkmann, A. (2003). Graphical knowledge display—mind mapping and concept mapping as efficient tools in mathematics education. *Mathematics Education Review, 16*(4), 35–48.

Carraher T. N., Carraher D. W., & Schliemann, A. D. (1985). Mathematics in the streets and in schools. *Br. J. Dev. Psychol., 3*, 21–29. https://doi.org/10.1111/j.2044-835X.1985.tb00951.x

Chiou, C. C. (2008). The effect of concept mapping on students' learning achievements and interests. *Innovations in Education and Teaching International, 45*(4), 375–387.

DeBellis, V. A., & Goldin, G. A. (2006). Affect and meta-affect in mathematical problem solving: A representational perspective. *Educational Studies in Mathematics, 63*(2), 131–147.

Mann, R. (2004). Gifted students with spatial strengths and sequential weaknesses: An overlooked and underidentified population. *Roeper Review, 27*(2), 91–96. https://doi.org/10.1080/02783190509554296

National Research Council (2000). *How people learn: Brain, mind, experience and school (expanded edition).*

Rizzolatti, G., Fogassi, L., & Gallase, V. (2007). Mirrors of the mind. In M. H. Immordino-Yang and K. Fischer (Eds.), *The Jossey-Bass Reader on the Brain and Learning* (pp. 12–19). Wiley & Sons.

Schoenfeld, A. H. (2011). *How we think: A theory of goal-oriented decision making and its educational applications.* Routledge.

Singer, F. M., Sheffield, L. J., Freiman, V., & Brandl, M. (2016). *Research on and activities for mathematically gifted students.* Springer Nature.

Sowell, E. J., Zeigler, A. J., Bergwall, L., & Cartwright, R. M. (1990). Identification and description of mathematically gifted students: A review of empirical research. *Gifted Child Quarterly, 34*(4), 147–154. https://doi.org/10.1177/001698629003400404

NOTES

Chapter Two

Empower Students to Learn Math

Empowering students to learn involves helping students take an active role in their learning!

Students learn best when they can make sense of information and try to figure things out themselves. This process allows them to take ownership of their learning and involves active thinking and learning from the student. John, a mathematician, shared that the Autonomous Learner Model used in his gifted program effectively supported his learning:

> I was enrolled in a gifted and talented program as a kid. In fact, the guy who taught it was a fella by the name of George Betts. He passed away, but he became a legend. He was my GT teacher in high school before he left to get his doctorate and become famous.
>
> The whole thing in gifted programs was to teach kids to be autonomous learners. He invented a model called the Autonomous Learner Model to teach kids to learn on their own, to learn by themselves, whatever interests them. It was the idea that kids learn to seek out the resources that they need.
>
> People could help them without being experts, right? If I had a kid who liked nuclear physics or whatever, but I knew nothing about it, I could help him. With the Autonomous Learner Model, a parent also knows how to help the kid develop that skill without having to worry about learning it themselves.

The Autonomous Learner Model by George Betts and Jolene Kercher focused on helping gifted children engage in self-directed learning. When learners are empowered to learn by themselves, the teacher or adults can ask questions, provide resources, and have the learners figure it out themselves. The teacher does not have to become an expert, but a facilitator, to help the students figure things out. All children, not only gifted children, can benefit when they can figure out how to learn something and be given the resources to do so.

Many of the individuals interviewed as part of the study for this book indicated how much they valued the *freedom to learn* things and *figure things out* for themselves, as opposed to having instruction prescribed in a step-by-step manner by the teacher. When math did not make sense, the people interviewed as part of the study shared that it became a negative learning experience. Typically, it is hard to make sense of math when taught through rote memorization of steps without understanding why the steps work.

All students benefit when they can actively participate in their learning. The Common Core Mathematics Standards (2010) for mathematical practice encourage students to think for themselves, be proactive in their learning, and engage in sense-making. Parents and teachers can support children by encouraging them to take *ownership* of their education and providing them with resources and support to figure things out. The idea is that the parent and the teacher do not have to be experts to help a child. Rather, they can help their children by providing them with resources to inspire their interests and figure things out.

SET A HIGH BAR FOR LEARNING

- Low expectations yield low results.

High-achieving students set a high bar for themselves and work toward doing well and getting an A. Setting a high bar also aligns with action. If a student is happy to get a C, they will achieve a C. If a student works on getting an A, they will reach that or at least get a B. When students set a low bar and skip topics to do the bare minimum, they set themselves up for failure. A mathematician explained that some students skip topics because they have calculated that they might pass the class with a C if they only study some topics. These students were happy getting a C. Therefore, encouraging students to set a higher bar for themselves and work toward that goal sets them up for success.

- Other distractions from the outside world may contribute to a lack of motivation.

Sometimes school may not be a priority because a student might be focusing on other things, such as the social activities taking place, while in school. Therefore, they may not set a high bar for themselves to get good grades.

Thus, the student might not get good grades. A mathematician named Keith, who got his undergraduate and master's degrees in math and pursued a doctorate in psychology, shared, "I was not that good at it in eighth grade. I got a D-minus in my algebra class."

When pressed further about why he got the D–, he replied that he simply did not care! However, he knew that he could do well. He had done well in elementary school. He was proud that he and his friend had won a math award. His parents were enormously proud of his accomplishment and made a big deal. This experience made him feel capable: "We won the math prize, and my parents made a big deal out of it, telling people about it and valuing it."

Fortunately, later, his interest and performance in math turned around when he went to college. The following section describes what successful mathematicians do when learning math and solving math problems. The important thing to note here is that sometimes grades don't reflect an individual's ability or potential to learn math. Performance in doing math can be impacted by other situations in a person's life. Therefore, teachers and parents must be aware of the child's potential and performance to provide support.

WHAT IS INVOLVED IN LEARNING MATH?

Seeing the "Big Picture"

Mathematicians interviewed shared that they see the big picture of what they are trying to accomplish and how the parts fit together when solving the problem. For example, consider the problem of sharing two pizzas divided equally into three parts. Students may get lost in figuring out how to slice a single pizza into three equal pieces and lose track of the original problem of splitting two pizzas.

John, the mathematician, shared:

> And I think that it is so important to make sure that they get the big picture! Yes, you should figure out those little details. But if you don't know "Where did this come from?" that should not get in the way of you understanding the concept. And so, I think parents can do a lot to remind their kids to keep their eye on the big picture.

The mathematician was asked what he meant by seeing the big picture. He emphasized how important it was for students to understand the "big mathematical idea" so that they could transfer what they learned into new situations.

What is this problem doing, and what are the techniques that they are using to solve this problem? Forget that you did not understand this calculation that went on within that problem. Do you understand *why* they started this way and *why* they got this number and answered the question this way? Does that make sense to you? Because that is the big meaning of the thing. The thing that you will try to carry over when you have a similar problem. You will not care where that came from when you have a similar problem. Because that too is a different problem!

Encourage students to monitor their progress of the whole problem and keep track of what they are doing. The goal is to break the problem-solving process into manageable chunks and keep track of the entire problem.

- What is the problem asking me to do?
- Which part am I working on?
- Does it make sense?
- Have I completed the whole problem?

Self-Diagnose Trouble Spots

The ability to self-diagnose is critical for being able to learn math efficiently. When solving math problems, an individual should identify what part of the problem makes sense and what part of the problem they find difficult. For example, a student should be able to say, "I get this part of the problem, but I am struggling with what to do next." When a student can self-diagnose and pinpoint exactly what they are struggling with, it is easier to ask for help, find resources, or ask a parent or teacher.

John, the mathematician who also teaches college math courses, shared that when a student can self-diagnose what they do not understand, he can work with that student. But if the student says they do not understand the entire problem, it is much harder to help the student.

> Kids like to come to their teacher and say, "I am lost; I do not get any of this!" It is so hard to work with a kid like that. If you have a kid explain, "I understood it here, but I don't know what happened after that," they're just missing those two up there, and they didn't see it. There are two down here; they do not know where those came from. They lose total confidence and think, "I am not good at math."

The mathematician shared that when students feel confident that they can figure things out, they develop competence. He pointed out that confidence is important in learning math.

Parents and teachers must empower students to think for themselves. Students need to solve problems with the *right challenge level to succeed.*

In other words, the student should have some background knowledge as a starting point, but the problem should also have a little challenge that forces the student to think and notice patterns. If the problem is too easy, not much thinking is involved. However, the student may not know where to start if the problem is too difficult.

PROBLEM-SOLVING STRATEGIES

Understand the Problem

Understanding the problem is the first part of solving any problem. The first part is to make sense of the problem. Albert Einstein said he spent 40 minutes trying to understand the problem and five minutes producing a solution. The goal is to encourage the student to understand the problem. Therefore, ask questions such as, "What is the problem you are trying to solve?" and "What is it asking you to do?" Have the student explain what the problem is asking. The following questions help students make sense of the problem:

- What information is given?
- What information is relevant?
- What is it asking you to do?

A good problem has a little bit of a challenge to it. Students should be encouraged to ask questions to solve problems and develop problem-solving skills. Questions such as "What is the problem?" and "What do you need to do to solve the problem?" help the learner produce a plan to solve the problem.

Start with What You Know and Work Your Way from There!

A wonderful way to tackle a challenging problem is to start with what they already know. Liz, an award-winning teacher and a parent of STEM graduates, shared:

> Take the scariness out of math. I teach my students to always start with what they know. And then you work your way out from there. And you plug in what you know to the other unknown parts. That is a life skill. You know where they can take it; they can tackle something in science or anything in life that seems overwhelming. It is just like a math problem where there is always a little piece you can do, but you know you can start with.

Create a Plan of Action to Solve the Problem

Once the problem has been identified and deconstructed, the student can devise a *plan* of action to solve the problem. Break the larger problem into smaller steps, focus on one part at a time, and keep going. Sean, a doctor who graduated from Yale medical school, provided the following example. He pointed out that understanding which part of the problem you are stuck in makes it easier to figure out a solution and get help.

He explained, "Think about a house. You do not have to rebuild the whole house if the light bulb does not turn on. You can try different things, such as changing the light bulb or checking the wires." This process is the same when solving a math problem. If someone gets stuck, it is helpful to know which part they did not understand. At this point, they can try to figure it out, look up resources, ask a parent, ask a teacher, and look at sample problems and class notes. They can keep trying things until they get it. Making mistakes is a natural part of problem-solving. Even mathematicians make mistakes when trying to solve complex problems. Understanding the error and how to fix it is an important part of the learning process.

Connect Math to a Real-World Context When Possible

The award-winning teacher who is also the mother of a STEM graduate shared that it helps when she connects what children are learning to real-world context.

> I tried to make math truly relevant. I have conversations with my kids when cooking, measuring to pour milk, and making Kool-Aid. It becomes so seamless that they do not realize that they are learning! You need to figure out how much sugar we will put in the two quarts of water. How many cups make two quarts? Math makes sense.

Visualize the Problem

Visualization of the problem situation and solutions also helps. When a problem can be related to real-world context, it is easier to visualize the math. For example, if you ask a student to divide ½ by 2, they only focus on the math procedure. They may invert and multiply and conclude that it is ¼. However, if they can visualize ½ as a pizza is cut into two equal pieces, then they will be able to visualize ¼ of the pizza, and the problem would make sense to a person just learning to divide fractions. A student can draw a picture, act out the problem, or build a model to visualize the problem. (See figure 2.1.)

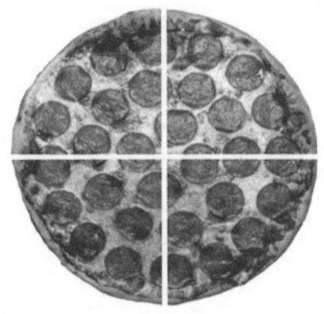

Figure 2.1. Visualizing the Problem Helps
Teruni Lamberg

TURNING OFF THE MATH BRAIN

- When students experience math without discovery and processing time to think, they lose motivation and interest in math. The impact can potentially devastate students' self-confidence and motivation to persist for a lifetime.
- Good teaching provides students with opportunities to think, explore, and discover. It builds confidence and self-esteem.

Many people often confess that they hated math and were never good at it. Interestingly, even some very smart, capable high achievers had negative experiences in elementary and middle school that made them lose interest and confidence in math despite the fact these individuals ended up being high-achieving students. For example, they did well enough in math to score well on the SATs or passed prerequisite math courses to get into good colleges through persistence because they were good students. However, their brains were turned off from loving math and they avoided pursuing STEM fields. Furthermore, they continued to believe that math was not an area of their strength and talent; therefore, they were not a "math person."

Interestingly, when these individuals were interviewed and reflected on their experiences, they suddenly became self-aware that they were capable and did well in math through perseverance. They could pinpoint *pivotal* moments that turned off the math brain from loving math. These experiences made them *lose confidence* in their abilities despite doing well enough to get into good colleges. Unfortunately, vivid memories of these *painful experiences* still linger after all these years.

The goal of this book is not to state that everyone needs to pursue careers in STEM. We need people in many disciplines, and certain people might have different talents and interests. *What is significant in this study is that pivotal events turned off the love of learning math even as early as elementary school.*

Vincent, a Harvard graduate who pursued human geography as an area of study, shared that he lost his confidence in math because he had difficulty understanding a concept in third grade. The teacher's *prescriptive approach* inhibited his ability to figure things out in a way that made sense to him. The teacher asked the students to follow steps to read elapsed time on a clock. This activity did not make sense to him, and he could not read the clock. The prescriptive approach and the pacing of the lesson were too fast. This learning experience was not only painful but embarrassing too! He felt everyone else other than him understood how to "tell time." When interviewed, he shared:

> I had a very distinct memory of third grade when we were learning about *time.* And I could not understand it at all! I just could not read the clock. I could not! It did not make sense, the way the teacher was teaching it. It got so bad that . . . for some reason, they put me in one of the remedial math classes that year. And I struggled with reading the clock. That was one of the concepts we reviewed, and I just remember feeling so bad that I had disappointed everybody.

Even though he did well enough in math to get into college and eventually graduate from Harvard, no one ever told him he was incapable of learning math. This internal struggle affected his entire life. The learning experiences associated with telling time affected his self-esteem because he wanted to be liked by his friends and feel included. He explained:

> So, I think it was more like my self-esteem. This inner dialogue of you just can't get it. You are not like, "How can you figure this out?" Instead, "You are not going to figure this out!" I cannot even read the clock like, "What a fool I must be!" Nobody was telling me these things! Nobody was using those words! It was all some inner dialogue. And still, to this day, I do not know where it comes from. Because it still creeps back in now and then.

This individual sadly concluded: "And so, you know, I think that *turned me off to math* when I began saying, 'I am just not a math person. You know, this is not easy for me. I can't do this!'"

This experience led to the development of a fixed mindset where the learner felt unable to learn how to read the time. Carol Dweck, author of *Mindset: The New Psychology of Success*, pointed out that people with fixed mindsets do not believe they can improve. They believe that their abilities cannot change. Therefore, the Harvard graduate's internal dialogue made him feel incompetent for the following reasons: being placed in a remedial class, not wanting to disappoint his parents, and needing to be liked and respected by his friends at his table. He felt humiliated because the other students knew the answers, and he did not.

*The critical point is that children may not have the **perspective** to see beyond those immediate negative experiences* that could potentially turn them off from enjoying learning math. Vincent shared that he would have had a more successful experience if the teacher had let him figure out how to read the clock on his own, instead of prescribing a method that did not make sense to him.

Well, I wish they would have let me just play with that Judy clock on my own to see how it works and manipulate it more on my own, rather than in a specific way. I wish they would just let me play with it on my own. You know, at a particular time, maybe it would have clicked better and faster.

He never shared this experience with his parents because he was so embarrassed. They could not intervene. He was silent in the group and did not advocate for himself to ask for help. When this individual reflected on this painful experience, he struggled because he was not allowed to make sense of the problem and figure things out for himself. Therefore, it had nothing to do with his ability. He realized that he could figure out the concept by himself if allowed to do so.

And over the years, I realized that I'm very good when you give me a book and say, "Okay, you have a week to figure out the concept and report back." And I get to figure out the concept on my own and report back to make sure that I understood it. I felt they never allowed that kind of time when we did group activities; no doubt they were fun.

Frustrating Situations in Learning Math
Have Psychological Implications

A teacher should extensively use formative assessment to inform teaching. Understanding student ability is particularly helpful when quiet kids in the class may not speak up if they don't understand something. The teacher has to be proactive in assessing students and intervening. For example, the teacher can walk around and observe what the students are doing, examine their work, and ask questions. The Harvard graduate explained that there were psychological implications for being unable to read the clock. He felt that everyone other than him was getting the concept. He explained that if he were allowed to explore the clock and figure out how to find elapsed time on his own, he would have been able to figure it out.

The Harvard graduate was not alone in having experiences of struggling in math that led to a loss of both confidence and a love of learning math. A Stanford graduate with a doctorate in the humanities also shared a similar experience. He, too, developed the identity of "not a math person" even though he did well enough in high school to get into college and pursue graduate school. Upon reflection, he realized that he was good at math and could even figure it out if needed. He even scored a 710 on the SATs. But he never developed a love of math.

> When it comes to math specifically, I did have a fixed mindset around math. But it was that I was not good at math. . . . But I did know I certainly did not conceive of my mathematical ability as something that could improve. And I do not think my parents did either. So, I do believe that they [parents], you know, probably do have a fixed-mindset mentality.

This Stanford graduate did not always hate math. He loved math in elementary school because it involved a discovery process.

> Where there was, you know, sort of like a puzzle. But when I was in third grade in math class, there was a discovery. . . . I remember learning long division, which was probably in third grade, and it just blew my mind! So cool that you could manipulate these numbers and create new things out of something. That was cool! I loved it up through fifth grade. And then I got to middle school and sixth grade; I don't know, it just changed.
>
> Maybe because there was less of that discovery, or it just got too hard for me. So, I think I did have some lousy teachers in middle school and high school in math who could not convey the joy of mathematics. And I just kind of lost interest!
>
> I went to college with no core requirements, so I never took a math class and never took a science class. I just walked away from that forever. Yes, I took it very seriously when I was there in the test room, and I figured it out. . . . The joy was never there!

These two examples illustrate that competent students can lose their *confidence* and *joy* in doing math when math taught is devoid of problem-solving and discovery.

- The brain turns off the love of learning math when there is a misalignment between the student's ability and the kind of instruction they receive.

When there is a mismatch between a student's background knowledge and the instruction provided, it can lead to ineffective learning. If the work is too easy, then students get bored quickly. A mathematician shared how bored she was doing endless pages of long division in third grade when she already knew how to do it. She did not see the point and dreaded the boring worksheets. The teacher used long division to punish the class if they misbehaved. If the level of instruction is too complex and students have gaps in their learning, it can lead to math anxiety. Don't let students be in a state of frustration because the content is difficult to comprehend.

A parent of a STEM graduate shared that he never wanted his kids to get frustrated. He strongly felt that if the lessons were too complicated, it would lead to frustration and demotivation. Therefore, he did everything to ensure that his kids asked him or their teacher for help if they were stuck. Therefore, parents and teachers should monitor that students are learning at their zone of proximal development and not at the frustration level. If the student is at a level of frustration, intervention would be needed to help the student.

The Zone of Proximal Development is a concept identified by Lev Vygotsky (1978) where the instruction matches the student's ability and can build on the student's prior knowledge under the guidance of a teacher or peers. Good teachers assess students and provide instruction that builds a bridge from students' prior knowledge (Lamberg et al., 2020). A discovery approach that integrates standards-based teaching builds students' confidence in their abilities. The book *Work Smarter, Not Harder: A Framework for Math Teaching and Learning* provides teachers a framework on how to teach using a discovery approach and integrates the Standards for Mathematical Practice from the Common Core Mathematics Standards.

NOBODY CARES IF YOU GET THE *WRONG* ANSWERS *FAST!*

- Give students time to think and process, instead of emphasizing speed, when they start learning.
- Once students become proficient and understand how to figure out the answers and notice patterns, they will develop speed with practice.

Ready, set, go! "Okay, you have 100 multiplication problems to complete in 10 minutes! See who can be the fastest!" Wesley sat in his seat and started on the first problem. He stared at all the worksheets filled with problems and thought,

"How am I going to get everything done in time?" He looked at the kid beside him, whose head was buried in the paper. This kid was writing fast. Seeing the speed at which the kid did his work made him more anxious. He started writing the answers and erasing them because of his mistakes. Out of frustration, he just started filling in the blanks without thinking about what he was doing. He hoped that he had guessed some of the answers correctly.

Keith, a mathematician who pursued psychology, shared:

Nobody cares if you get the wrong answer fast! And that's a guiding principle of a good math environment. That is not reinforced in standard pedagogy. Standard pedagogy is that I give you a worksheet with 50 problems on it. Get it done as fast as you can! Right, everything is fast, fast, fast, fast, fast!!! . . . Nobody is going to care if you get the wrong answer fast! People will be happy to pay if you get the correct answer, which takes a long time.

Learning math requires time to *think* and process. As individuals practice, they get the answer correct at a faster speed. If the individual focuses on getting the right solutions quickly while still struggling to learn, it sets an individual up for failure. Speed does not equal learning. Instead, taking time to understand how to solve the problem and the underlying math concepts helps. With practice, students will become more efficient and develop speed in solving problems. What is interesting to note is that even though standardized tests focus on speed, mathematicians spend months or even years trying to work out a proof (Sternberg, 1996).

SUMMARY: EMPOWERING STUDENTS TO LEARN

Empowering students to learn involves having students actively engage in solving problems and focus on learning big mathematical ideas. They need to see how the math concepts are connected so that they remember the math concepts better. The end goal is not the right answer to the problem. Rather, they learn how to solve the problem and use what they learned to solve other types of problems. Teachers can support students to engage in problem-solving.

Table 2.1 outlines strategies based on the Common Core Mathematics Standards (2010) to support children in solving math problems and making sense of what they are learning. Table 2.2 lists conditions that turn off the math brain.

Strategies to Motivate Students
• Support students to solve math problems by themselves. Instruction should be built on student ability. Encourage students to do their best!
• Help students set a high bar for themselves based on their ability. Figure out their strengths and weaknesses and build on assessment of their abilities. Encourage students to do their best!
• Observe and talk to students if they seem unmotivated to figure out the underlying cause so that it can be addressed.

Table 2.1. Problem-Solving Strategies

- Understand the problem.
- See the big picture of the problem they are trying to solve.
  - Ask what the problem is asking you to do.
  - What information is given and relevant to coming up with a solution?
    - Draw a picture or write an equation.
    - Figure out what operation to use.
What do you have to do to solve the problem?

- Start with what you know! Ask, "What do you know how to do?"
- Plan: Break the problem down into smaller parts and put it together at the end.

- Solve the problem.
  - Work on parts at a time in a logical sequence.
  - Some strategies when problem-solving:
    - Draw a picture or use manipulatives to visualize the problem.
    - Connect to a real-world context if that helps.
    - Write a number sentence that connects to the models.
    - Build on what you know.
    - Make connections between what you know and new information.
    - Ask yourself: "Does it make sense?"
    - Explain your thinking.

- Check your answer.
  - Make sure calculations are correct.
  - Answer the original question and combine all the parts working together.
  - Identify the part that did not make sense or is a trouble spot.

- If stuck, try to figure out what you did not understand.
  - Look at a sample problem in the textbook or notes and try to figure it out.
  - Look up a YouTube video.
  - Ask a parent, teacher, sibling, or someone else for help.
  - Figure out *why* and *which* part you did not understand so that if you encounter it again, you can solve it.

Study plan:
- Solve a similar problem and see if you understand it.
- Do practice problems; repetition is needed to get good at it.
- Write down the "big ideas" and formulas for future reference and refer to this sheet.

Table 2.2. Conditions That Turn Off the Math Brain

Out-of-School Environment	*Classroom Environment*
• An identity that "I am not a math person. I am not capable of learning math." • Parents state that they were not good at math too, so not to worry. • A parent does the homework for the child. • A child has too much homework and not enough time to process.	• Learning math through rote memorization without understanding what the learner is doing. • Students are not allowed to explore solving math problems in a way that makes sense to them. • Math instruction is either too difficult or too easy. • Focus is on speed, not understanding.

BIBLIOGRAPHY

Anghileri, J. (2006). Scaffolding practices enhance mathematics learning. *Journal of Mathematics Teacher Education, 9*(1), 33–52. https://doi.org/10.1007/s10857-006-9005-9

Betts, G. T. (1985). *Autonomous Learner Model for the Gifted and Talented.* Autonomous Learning Publications and Specialists.

Boaler, J. (2002). *Experiencing school mathematics: Traditional and reform approaches to teaching and their impact on student learning.* Routledge.

Bransford, J. D., Brown, A. L., & Cocking, R. R. (2000). *How people learn* (Vol. 11). National Academy Press.

Carraher T. N., Carraher D. W., & Schliemann A. D. (1985). Mathematics in the streets and in schools. *Br. J. Dev. Psychol. 3*, 21–29. 10.1111/j.2044-835X.1985.tb00951.x

Hodge, L. (2008). Student roles and mathematical competence in two contrasting elementary classes. *Mathematics Education Research Journal, 20*(1), 32–51.

Lamberg, T., Gillette-Koyen, L., & Moss, D. (2020). Supporting teachers to use formative assessment for adaptive decision making. *Mathematics Teacher Educator, 8*, 37.

Masingila, J. O., Davidenko, S., & Prus-Wisniowska, E. (1996). Mathematics learning and practice in and out of school: A framework for connecting these experiences. *Educational Studies in Mathematics 31*, 175–200.

National Governors Association Center for Best Practices, Council of Chief State School Officers. (2010). *Common core state standards: Math.*

Sternberg, R. J. (1996). What is mathematical thinking? In R. J. Sternberg & T. Ben-Zeev (Eds.), *The nature of mathematical thinking* (pp. 303–318). Lawrence Erlbaum.

Vygotsky, L. S., & Cole, M. (1978). *Mind in society: Development of higher psychological processes.* Harvard University Press.

Chapter Three

Everyone Benefits When Parents and Teachers Communicate Effectively

We have standardized tests and a standardized curriculum, but we do not have standardized children! Each child is unique, and we need to know their strengths and weaknesses to help them!

- When parents and teachers communicate effectively, life becomes easier for everybody!

Jamie gathered her courage and walked up the elementary school steps to talk to Ms. Jones. Her heart pounded as she slowly climbed the stairs and nervously grabbed her notebook. Words raced through her mind as she thought about what to say. "Do I tell Ms. Jones that she gave too much homework that Wesley did not understand? What if Ms. Jones thinks I am a terrible parent?" The last parent-teacher conference was not a positive experience.

Ms. Jones, too, was a little apprehensive that Jamie wanted to meet with her for another conference. She sat at her desk and put her hands on her head, trying to calm herself. The last parent she met accused her of being a lousy teacher and reported her to the principal. This encounter left her a little nervous as well. "What does Jamie want to discuss? What should I tell Jamie?" It had been a long day teaching, and she was exhausted.

The principal had spoken to Ms. Jones about improving her low test scores. This discussion added to her stress level. "If only parents could help their children at home, life would be much easier," she thought. "I cannot do everything by myself. I have over 28 children in my class. It is so hard trying to give everyone individualized support. I work so hard," she told herself. "My parents have no idea what I have been going through all day." The classroom door opened, and in walked Jamie. Ms. Jones met her at the door and greeted her. They both sat down at the corner table in the classroom.

27

What happens next can build a bridge for a meaningful relationship that mutually benefits each other. On the other hand, it could create a barrier to communication and a hostile relationship.

• Communicate with the teacher first and have realistic expectations.

The first line of communication must be with the teacher instead of the principal or the school board. Teachers get burned out when parents push too hard and have unrealistic expectations. It is much better to communicate and have realistic expectations of teachers. A parent whose child is a Harvard graduate also serves on the school board and had additional insights into parent-teacher communication. She pointed out that "you catch more flies with honey."

> I see a lot of people get unpleasant with the schools if they do not get their way. . . . Honestly, I think just talk to the teacher and have good, open communication. It is easier to do so with elementary teachers because you usually just have one teacher for the whole year. So, you can kind of establish a relationship, but acknowledge that it is kind of a hard job to do. But I think *communication* is the key.

Parents find it useful when teachers communicate honestly about the strengths and weaknesses of their children's abilities in math. A parent shared that it is useful when the teacher communicates honestly about the child's strengths and weaknesses without sugarcoating information. Knowing what parents can precisely do to help is useful. Specific feedback—such as, "Your child is struggling with the division of fractions"—provides the parent with something concrete they can work on or use to find resources to help the child.

> Ms. Jones thanked Jamie for taking the time to talk to her. Jamie thanked Ms. Jones for her time and said she appreciated everything she did for Wesley. She asked Ms. Jones what she could do to help at home. Ms. Jones pulled out Wesley's math binder, shared his work with Jamie, and pointed out the areas of both strength and struggle. Jamie shared: "Ms. Jones, I am unfamiliar with this new way of teaching math. This is not how I learned math. Do you have any suggestions, such as some videos or books you can recommend, to help Wesley?"

PARENTS HAVE INSIGHTS INTO THE STRENGTHS AND WEAKNESSES OF THEIR CHILDREN

Parents spend a lot of time with their children and have great insights into their children's capabilities. When a parent understands what a child should learn, it is easier to support them effectively at home. Furthermore, parents' insights about their children's learning styles, strengths, and weaknesses help teachers provide more targeted support. Parent-teacher conferences are useful avenues for communication and sharing of information. The goal is to develop a positive relationship between parents and teachers to build a partnership of support.

Parents Should Advocate for Their Children's Needs

Parents understand a child's unique talents or needs, which is helpful for teachers. Many parents can advocate for their child's needs by providing targeted support. For example, one parent shared that their child had difficulty copying math problems from the board due to handwriting issues. The parent was able to advocate that the child could take digital photos of the equations on the board, as opposed to handwritten notes. The child could focus on the lecture instead of getting lost in notetaking.

A child may need additional resources and support that the school district can provide—for example, support for a learning disability such as ADHD (attention deficit hyperactivity disorder). If a child has ADHD, it does not mean that the child needs to be placed in a lower-level math class. The child might need some additional support to stay focused and engaged. Many individuals have succeeded with ADHD, become mathematicians, or pursued higher education. One of the mathematicians interviewed said that he had ADHD and his ability to hyper-focus on math became an advantage.

> When I was in fourth or fifth grade, my mom talked to me and said, "You know that your teachers are not happy with you; you are always looking up at the board. You do not seem to know your cursive letters, and you are always looking up to see how they are made. And you just do not put much time into your work, and you do not seem to take much care." I remember just kind of turning a corner in sixth grade. I got super careful and precise, became, like, a super-type-A personality, and did everything to the best of my ability. And then, everybody left me alone.
>
> I have had ADHD all my life. So, as a kid, I was diagnosed with it . . . but it turned out that when I did mathematics, I loved it and was good at it. And I later realized this, even if you have ADHD: if you do something you love, you lock into the zone, and you get in the zone. It does not matter what disease; you can just perform incredibly well.

Betty, the parent of a Harvard graduate, shared that she had a gifted child who was isolated from the rest of the class to get individualized, challenging work that she could do alone. This isolation harmed the child socially and emotionally because she did not want to be singled out.

At one time, they had her sit at a desk in the classroom doing a unique indepen-dent study while all the other students were doing a project. . . . All by herself, and she hated it. I am sure they had good intentions, but it caused her to feel alienated, in a way she did not like. She did not like being separated from her classmates. And I am sure it did not help personal interaction at that age to be, you know, sorted out. And you know, the smart kids are going to go over to independent study. . . . She did not want to be classified like that.

When a student feels safe and supported in the classroom, the student is more likely to be more engaged in the lessons. The parent shared that she handled it by empathizing with the child and talking to the teacher to work out a solution. The teacher may not have realized that isolating the child made her feel alienated from the rest of the class. The teacher may have thought she was trying to help the child by separating her from the group and letting her work independently, without realizing its emotional effects on the student.

BE PRESENT IN A CHILD'S EDUCATION

• Parents must frequently communicate with their children and teachers to understand what the child is struggling with and to provide targeted support.

Many parents do not meet with the teacher unless it is a concern that they feel is affecting their child. Therefore, teachers should listen to parents' concerns. Paying attention to a child's learning needs is helpful. A parent can observe if a child is focused or paying attention when doing homework. Interestingly, most of the individuals interviewed shared that the parents were available for help when needed, and the child asked for help.

A mathematician interviewed said she knew she could always rely on her dad, an engineer, to help her understand math. They had a close relationship. The bond of love and support between a child and their parents helped. A parent can also observe a child's frustration and intervene if necessary.

I was lucky because when I had any of those problems in my classes, I would go home to my dad. He would help me. I mean, he helped me a lot, yeah, to make it click in my brain, and we thought a lot alike, so that was very helpful.

The reality is that not every parent knows how to support their child in understanding math. If a parent can figure out what the child needs, they can provide the right support. There are other things a parent can do, such as ask-ing the child what they are struggling with and giving the child the resources to help.

The bottom line is that a parent cannot always control what happens in the classroom. However, a parent can decide how to support their child at home. This means figuring out how to keep things manageable while juggling parenting responsibilities such as cooking dinner. Similarly, a teacher can observe the child in a learning environment and provide the parent with insights about the child. The teacher can offer parents suggestions for providing support at home. But the teacher can only control what happens in the classroom.

IDEAS FOR TEACHERS: THE PARENT-TEACHER CONFERENCE

Schools typically have parent-teacher conferences. Teachers must think about how to structure the time when they meet with parents. Most schools hold back-to-school nights at the beginning of the school year and parent-teacher conferences with individual parents during the school year.

A teacher must plan the back-to-school night for parents and create a welcoming atmosphere where parents feel welcome to visit the classroom and ask questions. During this back-to-school night, the teacher can share the expectations for teaching math (and other subject areas) and suggestions for how parents can help their children and communicate with the teacher. How the teacher greets and welcomes the parents sets the tone for the rest of the year.

Teachers realize that many parents are not trained to instruct their children. Therefore, they need to have realistic expectations of what parents can do to support their children at home. The following can help teachers prepare for back-to-school night.

Parent-Teacher Conference
Goal: Understand Teacher Expectations Questions to Ask the Child's Teacher • What math is my child expected to know? • Do you use a pacing guide? • Is there a digital platform where you posts assignments? • What are the procedures for assigning homework and turning it in? • What are the policies for makeup work? • How can we best communicate with you (e.g., through e-mail)?

Back to School Night: Suggestions for Teachers
to Talk to the Group of Parents

- Get to know the parents and have them share something special about the child.
- Create an inviting environment by greeting the parents and making them feel welcome.
- Start by having parents share something unique about their children.
- Share curriculum, calendars, and resources that will be used. In other words, what should a child learn during the school year, and how can parents help?
- Create a handout with info for parents and what they can do to help, such as:
  - Calendar of topics: outline what will be covered during the academic year
  - Textbook (physical or digital resources used)
  - Homework policy, due dates (better to have weekly homework), when to turn in work), and makeup policies
  - Class website (digital resources): helpful for sharing handouts, makeup work, and homework help
  - Class communication and teacher contact information

Individual Parent-Teacher Conferences

Individual parent-teacher conferences provide excellent opportunities for the parent and teacher to communicate about the strengths and needs of the individual child. A teacher can create a list of strengths and growth areas for the child to discuss with the parent. A written list is a helpful conversation starter and allows the teacher to reflect before meeting with the parent.

Suggestions for Teachers to Plan for Individual Parent-Teacher Conferences

- Strengths
  - How is the student performing in math?
  - What are some of the strengths observed in class participation?
- Areas of Growth
  - What is the child struggling with mathematically? Parents need specific details on what they can do to help their children. Provide some resources or suggestions for the parent to help their child.
- Questions and Comments from Parents
  - Have parents share their questions or concerns.
  - Write down suggestions and take notes for future reference.

Evidence of student performance, such as test papers, test scores, and student work samples, helps parents understand performance and needs. Teachers can create individual student folders and a bulleted list of talking points of strengths and weaknesses to discuss with parents. On the day of the conference, organize folders based on the parent-teacher conference schedule.

The teacher and parents can set a tone for collaboration and mutual support. Conducting a parent-teacher conference can be stressful for the teacher and the parents. Therefore, creating a relationship of *trust* helps eliminate the stress associated with conducting the meeting. The focus should be on how the partnership will benefit the child. The teacher can start the conversation by sharing something positive and then sharing "areas of growth" (as opposed to making it punitive, like "areas of deficit"). The parents can also share what the student is doing well and the challenges they notice at home.

WHAT ARE SOME REALISTIC EXPECTATIONS A TEACHER SHOULD HAVE OF A PARENT?

A parent does not have to be trained as a teacher to support their child. Therefore, it would be unrealistic to expect that all parents know how to teach math content at home. Furthermore, some parents work long hours and may not have the time to spend with their children. Therefore, a teacher must be thoughtful about what kind of work and support they can expect at home. The amount of homework should be reasonable and meaningful. It should also be something the child can work on independently. Sending home resources if the student is stuck, such as notes, links to videos, and manipulatives, is helpful.

CHOOSE THE RIGHT SCHOOL ENVIRONMENT FOR SUCCESS

- Just as plants need water and sunshine to grow, children need inspiring, caring teachers who focus on conceptual understanding to spark the math brain! The classroom environment matters!

The right classroom environment can set a child up for success or failure. The reality is that not every math teacher will inspire and care about each child as an individual. Many parents and students interviewed said that parents actively advocated for getting their children into classes that were aligned with their needs. Parents would volunteer in the school or ask other parents about teachers in order to decide which classroom to place their child in. The reality is that many parents work and may not have the time to volunteer.

If the environment is not a good fit, it might be beneficial to help the child navigate the environment or place the child in a classroom set up for success to meet the child's needs. Talking to other parents, school administrators, or teachers might be helpful. A mathematician shared a story of his parents, who were not educated but wanted a better life for him. They did not have much money and barely could afford to put food on the table.

The mother ensured that he went to good schools and kept changing schools until she found the right fit. When the mathematician reflected, he shared that he did well in math in high school because he had great teachers who cared and focused on conceptual understanding. Finding the right environment set him up for success.

The advice given by several people interviewed is to recognize when a child is in an environment that is not conducive to learning and take action to help the child. Teachers, peers, friends, and family can potentially become a child's support system.

PRIDE MATTERS: TREAT TEACHERS AND THE MATH TEXTBOOKS WITH RESPECT

Messages that children hear about the teachers and math learning impact a child's motivation. Keith, a mathematician, psychologist, and parent, shared that math pedagogy has changed over the years. He explained that current math teaching in schools might differ from how parents learned math. Therefore, parents should be careful not to give the message that the teacher and textbooks are dumb.

> And there is an intrinsic emotional reaction from the parents or the tutor, and sometimes, even the teacher calls the textbook stupid. If you say that your textbook is dumb, your teacher is dumb, and this is the wrong way to do it, then a child is being taught not to be proud of what is being asked of them. He suggested that parents should say something like: "It is different from how I learned it, but this is awesome. This is cool; your textbooks are amazing." You want them to be proud of what they're learning.

Treating teachers and textbooks with respect is important to show children how to take pride in their actions and be thankful for their experiences. If the environment is not conducive to the child, communicate with the teacher or school or even make changes in the environment.

WAYS PARENTS AND TEACHERS CAN
COLLABORATE TO RAISE AWARENESS

When parents are involved in their child's education, children perform better (Hoover-Dempsey & Sandler, 1995; Sheldon et al., 2010). One way that teachers can engage parents is to conduct Math Family Nights. Lynn Liao Hodge and Michael Lawson (2018) provide guidelines for teachers and schools to run Math Family Nights. They suggest having an inviting space, providing choices for participation, and having workstations where there are math materials and clear directions to have math ideas come to life.

An interesting finding from the research is that how children thought of themselves as math learners was related more to how their parents perceived their ability than the actual grades they earned (Sheldon et al., 2010). Parents have a strong influence in how they support their children. Therefore, teachers can help parents be involved in their children's education. Parents can take an active role in supporting their children.

Students don't often connect math used in their home and formal school math. Teachers can get to know the students and their home environments. One of the things teachers can do is incorporate things from the home environment into the lessons. For example, Gonzalez and colleagues (2001) gave the example of Señora Mira, who naturally used sophisticated math when she was sewing. This activity involved measuring, geometry, and mathematical calculations.

Students can see connections between the math naturally used in the home environment and the formal math they learn at school. Gonzalez et al. (2001) point out that not all communities, such as African American, Mexican, and Native American communities, have the same type of Eurocentric knowledge of math. Therefore, teachers can incorporate math from the home environment so students can make connections.

SUMMARY: EVERYONE BENEFITS WHEN PARENTS
AND TEACHERS COMMUNICATE EFFECTIVELY

Parent and teacher communication is important for developing a collaborative relationship to support the child. Therefore, the relationship is positive and productive. Parents have insights into the strengths and weaknesses of their children. Teachers observe children in academic settings and should provide factual information about the child's strengths and weaknesses so that parents can provide targeted support. The relationship should be based on mutual respect. See table 3.1.

Table 3.1.   Characteristics of Effective Parent-Teacher Communication

* Positive
* Collaborative
* Effective (supports student learning and success)

Suggestions for Parents	*Suggestions for Teachers*
• Keep it positive. • Have realistic expectations. • Share observations and concerns. • Advocate for accommodation if needed. • Be present with your child's education; observe what they are doing. • Change the environment or find the right setting to meet the needs of the child.	• Communicate strengths and weaknesses in a positive way. • Share specific and concrete data and provide parents with resources. • Create a regular communication system. • See chapter 3 for ideas for parent-teacher conferences and back-to-school night.

BIBLIOGRAPHY

Gonzalez, N., Andrade, R., Civil, M., & Moll, L. (2001). Bridging funds of distributed knowledge: Creating zones of practices in mathematics. *Journal of Education for Students Placed at Risk, 6,* 115–132.

Hodge, L., & Lawson, M. (2018). Strengthening partnerships through family math nights. *Mathematics Teaching in the Middle School, 23*(5), 284–287. https://doi.org/10.5951/mathteacmiddscho.23.5.0284

Hoover-Dempsey, Kathleen V., & Sandler, H. M. (1995). Parent Involvement in children's education: Why does it make a difference?" *Teachers College Record, 97*(2): 310–331.

Sheldon, Epstein, J. L., & Galindo, C. L. (2010). Not just numbers: Creating a partnership climate to improve math proficiency in schools. *Leadership and policy in schools, 9*(1), 27–48. https://doi.org/10.1080/15700760802702548

Chapter Four

Decoding the Math Standards
What Should My Child Be Learning?

Standards describe what students should learn developmentally at each
grade level.

THE COMMON CORE MATHEMATICS STANDARDS

- The Common Core Mathematics Standards outline what math concepts
  students should learn at each grade level and how these concepts progress
  across the grade levels.

The Common Core Mathematics Standards outline the mathematics students
should learn at each grade level. Many states have adopted the Common Core
standards or created a version of their state standards. Parents should know
the mathematics standards used in their children's school and where they can
find a copy of the standards. The Common Core standards outline a set of
practices they call "Mathematical Practice" on how students should experience math learning.

The Standards for Mathematical Practice incorporate a problem-solving
approach to learning math that involves looking for patterns, thinking about
the problem by creating models, and engaging in sense-making and logical
argumentation. Communicating thinking to others and being able to justify
an answer by providing evidence is an integral part of the standards. Unfortunately, many students get the love of learning math in their brains turned off
because they do not experience learning math this way!

Understanding what a child should know as they enter a grade is helpful
to parents and teachers. Parents and teachers can decide if the student needs

additional help beyond class instruction. A parent can determine if they need to get a tutor or provide additional support.

UNDERSTAND WHICH MATH CURRICULUM IS BEING USED

Supporting children at home is easier if the parent knows which curriculum is being used in the classroom. Many parents complain that they have no idea what homework is being assigned and where to find copies. There are lots of online resources available that children could use to help them figure out their homework too. Many textbook companies are creating digital resources, and teachers may have class web pages with resources. Some teachers do not send textbooks home. Therefore, having a copy of the book at home or understanding how to locate digital resources is helpful if the child has missing assignments or needs additional help.

WHAT DOES THE ASSESSMENT DATA SAY ABOUT MY CHILD'S STRENGTHS AND WEAKNESSES?

Teachers collect many forms of assessment data, such as state test data, district assessments, unit tests from curriculum units, and informal assessments. The important part of the assessment is not simply collecting the data but using it meaningfully to help the student.

Effective teachers plan and adapt lessons that build on students' prior knowledge. Good teachers use this data to design lessons that help students learn math concepts. The teacher can also use the data to advocate for the child if they need additional support because they have gaps in their learning, such as through interventions by special education teachers or gifted teachers. Understanding how the child is doing is also beneficial for the parent to advocate for and support their child. A parent can decide how to help their child by providing direct support, finding resources, or even hiring a tutor.

A caution about relying only on test data is that it is only one form of measurement of a child's capabilities. A lot of times, grades may not tell the whole story of a child. For example, the child may not have felt well when taking an assessment. A child who aced the test might have memorized the information for the examination and forgotten all the information a week later. Nevertheless, test data does give you insightful information about your child's strengths and weaknesses.

If a child gets bad grades, realize that those grades may not define a child's ability. Ying, a mathematician, shared that in third and fourth grade he was so focused on playing soccer that his grades suffered because he played three soccer games daily—before, during lunch, and after school. Either the child did not prioritize grades, or how the material was presented was not optimal.

In addition, he shared insight that he had difficulty learning the material because it involved a lot of memorization, and that he was not focusing on understanding. He struggled with this mode of learning and thought he might not be capable of learning this material. However, later during high school, things clicked for him. He started to see connections and the beauty of mathematics in high school.

Teachers can help by providing children and parents with resources and suggestions to help the students. For example, one teacher made short videos for parents on her phone and sent them as an e-mail, created a website with short videos, or provided access to videos on other websites.

IT IS OKAY IF THE PARENTS DO NOT KNOW MATH—TEACHERS ARE THERE TO HELP!

Teachers can provide parents with resources to help their children and reassure the parents that their children are capable of learning math. The parent of a STEM graduate and an award-winning teacher in a small rural school district shared:

> I have succeeded at teaching because I can take abstract things and make them concrete. And I try to do that visually. . . . Sometimes, I will take a piece of paper at a [parent-teacher] conference, and I'll draw out an explanation of something mathematically, and they will go, "Oh, oh, I see. Okay, I never even understood that when I was a kid!" And then they go home and are excited because I took the mystery out of it a little bit. So, I try to just make everything simple and relatable to stuff that they already know how to access.
>
> I also tell the parents that if they do not know something, call me up or send their child back to school and have them ask me. Some parents get confused and just give up and say, "Oh, that's too hard. We can't do that!" So, I try to encourage them. First, it is okay that they do not know because they learned it differently when they were young. So, I reassure them and give them the confidence that they can help their students.

See table 4.1 for suggestions on reflecting on students' needs and strengths.

Table 4.1. Reflecting on a Student's Needs and Strengths

Performance	Action
Is my child performing at grade level?	• Monitor grades to make sure the student is keeping up with the work. • Communicate with the child to ensure they are not struggling with a specific topic. Help the child develop skills to diagnose what they are struggling with, ask for help, or figure out resources. • Provide targeted support. If stuck, help them look for a video, and ask the child to advocate for themselves. Ask the teacher.
Is my child performing below grade level?	• Evaluate why the child is below grade level. Is there a learning disability? • Address any learning needs to figure out why. • Advocate for accommodations such as an IEP (Individualized Education Program) and 504. • Get a tutor. Tutors are available from tutoring services and online. Local colleges have tutors. If a parent does not have the resources to get a tutor, try asking a family member or friend to help. Ask the teacher for suggestions or ideas. Explore online resources.
Is my child not being challenged?	• Advocate for gifted programs.

TEACHER FEEDBACK TO CHILDREN AND PARENTS SHOULD BE CONSTRUCTIVE

Teachers should be mindful of the feedback they give parents and students about the capabilities of a child. If negative feedback is provided where the child feels incapable, it sets the child up for failure. The feedback should be helpful so that the student can learn from mistakes and do better next time. Inflexibility in grading where a child feels punished is not beneficial. When interviewed, Jeff, the high school class valedictorian with a master's degree in engineering, shared:

> Okay, so the teacher gives me zero. Well, I am going to keep doing what I am doing because I believe in myself. Poor feedback can be really disheartening to some people, and that is tough. I think, at some point, having faith in yourself is a big thing. And I wish that I had the answer to how to instill that in every single student. But you know, having someone to guide you and just having faith in yourself to get through those criticisms is important.
> Giving feedback that is disheartening. . . . You're basically ruining someone's dreams and their aspirations by providing feedback like, "You cannot do this.

This is all wrong! You don't have this skill." That is so sad to me, and I just go back to that support system.

Parents can help guide the child through negative feedback and build confidence. Teachers should focus on what the student is doing well and specifically what they can work on to support their learning. Addressing a child as incapable or a failure can have devastating results.

DO WE NEED A LOT OF HOMEWORK TO LEARN MATH?

Homework is helpful for a child to practice what they learned in class and reinforce concepts that they know. It can be a valuable tool to support learning when designed and implemented thoughtfully (Carr, 2013). The real question is "How much homework is necessary to support learning?" Research on the effectiveness of homework is mixed. If the assignment is not meaningful and is busy work, it may demotivate students from doing the work.

Homework should be an opportunity for students to make mistakes and practice. Homework should focus on the students practicing what they learned and thinking about what they understand and do not understand so they can get clarification in class. If homework focuses on only getting the correct answer without grasping the underlying math concepts, then homework can be very stressful for the child.

The amount of homework assigned must be reasonable. If a child cannot keep up with the number of tasks assigned each night, they may not do the work. The student may lose motivation to do math homework. Parents should communicate with the teacher if too much homework is assigned or a child is struggling with homework.

Sometimes, students may not do homework because they don't see the relevance in doing the work. A Stanford graduate said he never did homework because he recognized that he was smart and could grasp things faster and do it at school. He got good grades in school. He explained that he coasted through elementary and high school without much work. He started becoming more serious when he visited a school where the students were affluent and had goals of going to college. He realized that he needed to pay attention to what he was doing. He started to take his grades seriously at Stanford to get into graduate school.

When I got to college, I started taking it seriously partly because I could tailor my learning and studying to what I wanted to study. But I just kind of remember getting the spark of I should study, what I find interesting, and I don't care about anything else. And then that kind of pushed me to be motivated when it came time to go to college.

Homework should be meaningful and reasonable. Children need time to unwind and explore other things at home. Nothing is worse than a child who is burned out from doing homework all night and not motivated in class to keep working. When children have the opportunity to play logic games, get physical exercise, and explore math in informal settings, it builds a foundation for learning more formal math and to see meaning and relevance in what they are learning.

One of the most important things about homework is that students must learn how to manage their time to complete the assignment. The ability to regulate their time, plan, and monitor their progress toward their goals is important for success. Parents help their children keep track of and monitor their progress by asking them to make a list and check off what is completed, which helps students become successful (Xu et al., 2014).

SUMMARY: DECODING THE MATH STANDARDS— WHAT SHOULD MY CHILD BE LEARNING?

The math standards outline what students should be learning at each grade level. Teachers and parents should clearly understand what students should learn at each grade level. Assessment data is useful for understanding if the student is on track, has gaps in their learning, or needs more challenging work.

Teachers and parents must advocate for students to ensure they are learning and that the instruction meets their needs. When assessments have identified specific things a child needs help in, it is easier to provide targeted assistance. When teachers and parents communicate and work together, student achievement increases. Parents don't need to know the math to provide support, and teachers are there to help. See table 4.2 for a list of questions that parents and teachers can ask.

Table 4.2.   Questions That Parents and Teachers Can Ask

Parents	*Teachers*
• What standards are the math standards?	• What are the standards and assessment data on how the child is performing?
• What should my child be learning?	• What are some suggestions on the kind of support the child needs?
• What resources are available?	• What kind of resources should I provide parents?
• What questions do I have for the teacher?	• How can I make homework meaningful and reasonable? (Students should be able to complete homework independently.)

BIBLIOGRAPHY

Carr, N. S. (2013). Increasing the effectiveness of homework for all learners in the inclusive classroom. *School Community Journal, 23*(1), 169–182.

Xu, J., Yuan, R., Xu, B., & Xu, M. (2014). Modeling students' time management in math homework. *Learning and Individual Differences, 34*, 33–42.

NOTES

Chapter Five

Math Homework

Getting It Done!

Every child is capable of learning if you build a bridge between what the child knows and new information!

Parents of high-achieving children create a home environment that is stress-free and supportive. The home environment supports children to be active participants in their learning, and parents play a supportive role if they are stuck or need help. When children feel they have control over their education, they are more motivated to take initiative and be self-motivated. When parents are too controlling and pressure a child, the dominating interactions lower intrinsic motivation and math achievement (Steinberg et al., 1991).

TIME MANAGEMENT AND MATH HOMEWORK

- The homework should be reasonable, meaningful, and age-appropriate.
- Do not drive away passion by assigning too much homework.

Teachers should be mindful of how much homework they send home. A parent shared that her 13-year-old son received a weekly homework package with 52 pages of problems. The student was overwhelmed and unmotivated by the sheer amount of homework. Furthermore, the homework assignment also stressed the parent to stay on top of the student.

Therefore, teachers should have realistic expectations for homework. The homework should be something that has been covered in class and the child can do independently. It needs to have the right amount of challenge, practice, and application of knowledge learned in class. Most of all, the workload should be reasonable and appropriate to the child's grade level.

Homework should be something that students should be able to do independently and without help. Students must be able to figure out what homework they are supposed to be doing and manage their time to get the work completed. Teachers should think about homework's purpose and how to support student learning (Cooper et al., 2006).

PARENT EXPECTATIONS OF THE IMPORTANCE OF HOMEWORK, AND COMPLETION MATTERS!

Parents and teachers should expect students to complete their homework and be held accountable for their work. All parents interviewed felt that students should get their homework done and be responsible for completing it. How parents supported their children in doing homework differed. Some parents left it up to their kids to decide when to complete it. Other parents expected homework to be done by a certain time. Students did the work and decided how to approach the homework. Ryan, the Stanford graduate, shared:

> We had to complete my homework for the next day. You know homework was always very important. There was time set aside to do homework, usually after dinner, before I could relax.

One overarching theme is that the parents allowed their children to do their homework without intervening and hovering over them. Parents monitored student learning by paying attention to grades and talking to their children to ensure things were going okay. The child was expected to ask for help if needed and would be provided support. However, it was expected of the child to be responsible. The key point is that parents had to figure out when to provide support and when to let the child struggle. The parent can provide help or resources for the child to problem-solve. The other times, when the parent notices that the child is frustrated and having difficulty, the parent can intervene. Parent attitude and high expectations can positively impact student performance, according to researchers (Chen & Stevenson, 1989).

CREATING CONDITIONS FOR SUCCESS

Parents and teachers can help students learn by helping them to study more efficiently and stay focused.

Time Management

Time management is crucial for students to handle stress and break up large tasks into manageable chunks. The ability to manage time leads to homework completion and academic success. In other words, it makes sense to work smarter, not harder! Therefore, time management skills become essential.

Researchers point out that time management is one of the most important parts of the ability to do well in school (Corno, 2004; Eilam & Ahron, 2003; Pintrich, 2004; Schunk, 2005; Zimmerman, 2008). High-achieving students can monitor their time and regulate themselves (Zimmerman & Martinez-Pons, 1990). The goal is to empower students to be independent, but support them as needed.

Setting Goals

Students need to set goals for themselves and monitor their progress in accomplishing them. Therefore, the student must identify what they need to learn and what work needs to be completed—for example, breaking down the homework into chunks and setting goals to accomplish tasks. Create a checklist of tasks.

Create a Study Schedule

Empowering a child to think about what they need to do to study, complete homework, and manage their time to get their work done teaches them time management skills. Having them complete a task checklist and share it with the parent can be helpful.

Design a Study Space

Ideally, a child needs to have a desk and a quiet, uncluttered space free of distractions to study, with access to materials and supplies. A place to keep school supplies that includes materials, access to textbooks, and so forth is helpful. A scientist interviewed shared that she had a dedicated study space with a desk. She treated this space as her "office." It created the mindset in her that she needs to get her work done, which is important.

Create a Structured Routine

It is helpful to have a predictable homework routine so that doing homework is automatic. Have a clock so that the student can monitor time. Some parents expected the children to do their work before anything else. Others allowed their children to figure out a time that consistently worked for them. The children were held accountable for getting their job done.

Monitor and Accountability

A child should be able to hold themselves responsible for setting goals, regulating their time, and monitoring their progress. A parent can help foster children to self-regulate by asking the child to be accountable for their improvement. Teacher feedback is also helpful for the student to self-regulate and manage their time.

Provide Feedback

Parents can provide helpful feedback to get their students to figure out what they should be doing or to problem-solve when they get stuck. Parents can communicate expectations to do well and make the learning experience enjoyable.

Motivation

Parents and teachers should have high expectations of students. Students need to see that homework is important for helping reinforce what they learned at school.

Types of parent interactions that are not helpful include:

- *Parents who do the homework for the child.* When a parent does the homework for the child, the child does not develop the skills to do the task themselves.
- *Giving students the answers*, not letting the students figure out the answers themselves.
- *Not letting the child initiate and ask for help.* Providing too much support makes the child rely on the parent to help with the homework.
- *Lack of routines.* In other words, the child does not have a home routine that includes doing homework.
- *Not holding the child accountable for getting their work done!*
- *Having low expectations for the child.* "I am not good at math, so I understand if you are not good at math too!"

GET RID OF THE CLUTTER! CREATE VISUALLY PLEASING MATH PROBLEMS AND SPACE TO THINK

When writing a paper, students are encouraged to create multiple rough drafts and present the final product in a polished format of being neatly written or typed. There is something satisfying about seeing a written paper in its final form. Most students feel a sense of pride when looking at the finished product. Keith, the mathematician who got his doctorate in psychology, shared how important it is to create visually pleasing final products when solving math problems. He explained that we expect elegantly written work in literacy, not math. When math problems are neatly written in a visually pleasing manner so that they are easy to read and understand, it develops a sense of pride. Turning in beautifully written work creates a sense of satisfaction!

> Math can get messy and visually messy. You know kids have bad handwriting. And you know they're forever erasing, and crossing off in math gets messy. And it can be confusing if you want to go back and check your answer. So, one of the things that I have come to really appreciate that is critical for a good math environment is tons and tons of scrap paper.

The purpose of creating visually pleasing math problems is so students can go back and look at the work to make sense of it. If the solutions are haphazard, it is much more difficult to rethink the strategies used to solve a problem. For rough drafts of math problems, see figure 5.1.

The observation that this mathematician made with his children is that there is inadequate space in textbook worksheets. The lack of space restricts the student because they must keep erasing when making a mistake, and the paper looks messy. A solution is to provide students with plenty of scrap paper so they can think about the problem and have the freedom to make mistakes, cross things out, write in the margins, and then transfer the elegant solution onto a clean sheet of paper.

> There is a science they call visual discomfort—that cluttered and disorganized things can actually induce anxiety and stress. There is a field of neuroaesthetics . . . like the characteristics of images, shapes, and forms that we find pleasing.

This mathematician also recommended giving children plenty of graph paper so they have straight lines. He explained:

> It tends to be like curved, clean lines more than jagged. . . . You see, art is designed to provoke a sense of agitation. That is very consistent with what a messy worksheet would look like. The crucial point here is that children need space to write their thoughts without constraints and copy their solutions to be visually pleasing.

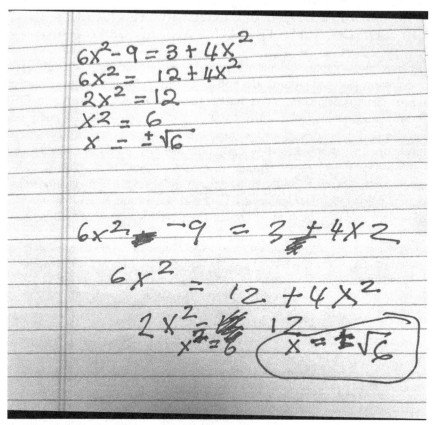

Figure 5.1.   Rough Drafts for Math Problems
Teruni Lamberg

- Get rid of the clutter in the physical space and make room to think!

Physical space can create a calm environment or be stressful if filled with clutter. Having a table clear of clutter can help a student to think because it is visually pleasing and uncluttered. Many individuals interviewed said they liked doing homework on the kitchen table because there was a large amount of space. Others had special desks dedicated to studying. Keith shared:

> I am going to have [my daughter] do her math homework. I cleared the desk. I have her work at the dining room table with as much space as she needs. There is a stack of scrap paper, and she has sharpened pencils.

LEARNING HOW TO STUDY

Learning how to study can transform a student's learning experience. Vincent, the Harvard graduate, shared that he took a class on learning how to study and that experience is transformational. He shared that many books and YouTube videos on study strategies are available.

It blew my mind because I thought studying meant just reading a book and memorizing things. And [the teacher] introduced me to a much more dynamic way of self-learning. She said, "If you are a visual learner, take lots of notes in class. I read and reread the text. Read it, you know, over a couple of days, draw pictures, and make a game, think of mnemonic devices." I did not know that those were things I could create or do.

I just never was aware that it completely changed how I learned. It changed my math learning completely. I was not doing well in my geometry class before I had that session with her. I flipped from a B to an A in that class. And I maintained straight As through math in high school since then, and a lot of credit goes to that teacher.

Study Tips for Math
1. *Listen* to understand.   • Focus on understanding math.   • Ask the teacher or another student to explain a part that does not make sense.   2. *Pay attention to "big ideas."*   • What formula do I need to remember?   • Do sample problem. If stuck, study the example. Ask yourself, "What is happening here?" Try to make sense of what you are doing.   3. *Do practice problems.*   • Try to do all of these. The more problems you try, the faster you become at solving the math problem, just like playing the piano. It will help you remember the patterns. It also makes you feel good when you do it.   • Check answers. If you missed a problem, figure out why.   • Doing math problems on a whiteboard helps students see the problem visually and lets kinesthetic learners use the body.   4. *Develop memory and speed.*   • Connect to prior knowledge.   • Practice repetition of formulas. Use flashcards, create a formula sheet, think of something funny, or connect to a real-world example if you can; for example, throwing a baseball to try to figure out the curve of the ball's flight.

PRACTICE MAKES PERFECT!

Practice is important in learning and making things stick in the brain. To get good at a sport, you must put in endless hours. If you want to play the piano, you must practice. When learning to paint, you must practice mastering techniques and skills. Math is no different. Betty, the Harvard graduate, shared that she took the calculus book home her senior year and practiced.

> The strongest one is calculus and senior year. . . . I struggled with the integral and took the textbook home.
>
> Because I did not understand what was happening in class, like working through problem examples, the teacher said she had never had somebody who just did practice examples on their own.

READING THE TEXT HELPS!

Many students enter college and gloss over the class texts. The ability to think about the textbook as a resource and a place they could go to seek answers when struggling is a great skill. Diligent students may take a math book home, look at the examples, and figure out how to solve the problems. John, the mathematician, shared that he found a lot of college kids did not read textbooks.

> I probably read the textbook a lot more than most kids do. And that was fine for me.
>
> But I do not think that is the case, of course. A whole lot of kids do not want to read the textbook, they do not want to read my answers to the problems, and they do not want to read anything. And that is something that I think is really holding kids back! They need to learn how to read at least a little bit to figure stuff out.

THE MIND AND BODY CONNECTION

Many students struggle with sitting and paying attention, while others can adapt to an environment where they must sit and do work. Some students need to be active and burn energy. Jose, the Stanford graduate, shared that doing something physical when he is engaged in a cognitive learning activity helps him. This combination allowed him to be a successful student at Stanford. He shared:

Suppose I am doing something physical while using my brain. My multitask was usually like physical activity, so maybe if you are trying to recite a poem while watching a documentary, that might be a divided thing with multitasking. But it's called multitasking.

Jose pointed out that swimming, biking, or running becomes automatic, and he does not have to use his brain. He is just moving his body. However, it helps his thinking and learning. He shared that what we do to students in school is unnatural. We expect students to sit for hours completing worksheets that do not have meaning. He contrasted that with how hunters and gathers learned in indigenous communities:

> They go out into the woods. And it is all about observation; they are just carefully observing and then go out and try things themselves.
>
> Yet in this factory model of education, you could do little tweaks here like, "Oh, we're going to mix it with multimodal learning and have some video." I mean, there are tweaks you can do to make them more bearable. But yeah, the physical classroom structure is just an appalling way to learn for most people.
>
> Take away the need to burn the energy, so you punish them by having them take recess to burn energy. So, you are just guaranteeing that they will suffer emotionally.

Some children need to be physically active to stay focused. Therefore, playing outside and participating in sports helps. Many of the successful students interviewed played sports, went outside, and socialized. Mitchell Nathan (2022) authored a book on embodied cognition. He points out that learning is not just limited to the brain. It involves the body too.

SUMMARY: CREATING AN ENVIRONMENT FOR LEARNING AND HOMEWORK

Learning how to study does not happen naturally. Children need to be exposed to strategies to explore. A home environment that is more conducive to studying and completing homework can be created. Teachers can support parents and students by assigning a meaningful and reasonable amount of homework to be completed independently. Teachers and parents can use technology to communicate and find resources to help the children when they are stuck (see table 5.1).

Table 5.1. Ideas for Parents and Teachers to Help Students Develop Study Skills

Ideas for Parents	Ideas for Teachers
• Encourage students to monitor their time to complete homework. • Parent expectations of homework completion matters. • Help students create study routines. • Create an uncluttered environment. • Teach children study skills. • Encourage practice. • Have children read the text. • Encourage children to exercise and take breaks when encountering mental fatigue.	• Focus on learning and understanding as opposed to test-taking skills. • Homework should be meaningful, reasonable, and age-appropriate. • Teach children study skills. • Have clear expectations. • Use technology to organize work so parents and teachers know what is expected and can find resources if a worksheet is missing.

BIBLIOGRAPHY

Boaler, J. (2005). The "psychological prison" from which they never escaped: The role of ability grouping in reproducing social class inequalities. *FORUM, 47*(2–3), 135–144. http://dx.doi.org/10.2304/forum.2005.47.2.2

Boaler, J. (2010). *The elephant in the classroom: Helping children learn and love maths.* Souvenir Press.

Chen, C., & Stevenson, H. W. (1989). Homework: A cross-cultural examination. *Child Development, 60,* 551–561.

Cooper, H., Robinson, J. C., & Patall, E. A. (2006). Does homework improve academic achievement? A synthesis of research, 1987–2003. *Review of Educational Research, 76*(1), 1–62. https://doi.org/10.3102/00346543076001001

Corno, L. (2004). Introduction to the special issue work habits and styles: Volition in education. *Teachers College Record, 106,* 1669–1694.

DiStefano, Michela, O'Brien, B., Storozuk, A., Ramirez, G., & Maloney, E. (2020). Exploring math anxious parents' emotional experience surrounding math homework-help. *International Journal of Educational Research 99,* 101526.

Dweck, C. S. (2006a). *Mindset: The new psychology of success.* Ballantine Books.

Eilam, B., & Aharon, I. (2003). Students planning in the process of self-regulated learning. *Contemporary Educational Psychology, 28,* 304–334.

Nathan, M. J. (2022). *Foundations of embodied learning: A paradigm for education.* Routledge.

Pintrich, P. R. (2004). A conceptual framework for assessing motivation and self-regulated learning in college students. *Educational Psychology Review, 16,* 385–407.

Retanal, F., Johnston, N. B., Di Lonardo Burr, S. M., Storozuk, A., DiStefano, M., & Maloney, E. A. (2021). Controlling-supportive homework help partially explains the relation between parents' math anxiety and children's math achievement. *Educ. Sci, 11,* 620. https://doi.org/ 10.3390/educsci11100620

Schunk, D. H. (2005). Self-regulated learning: The educational legacy of Paul R. Pintrich. *Educational Psychologist, 40*, 85–94.

Steinberg, L., Mounts, N., Lamborn, S., & Dornbusch, S. (1991). Authoritative parenting and adolescent adjustment across various ecological niches. *Journal of Research on Adolescence, 1*, 19–36.

Zimmerman, B. J. (2008). Investigating self-regulation and motivation: Historical background, methodological developments, and future prospects. *American Educational Research Journal, 45*(1): 166–183. https://doi.org/10.3102/0002831207312909

Zimmerman, B. J., & Martinez-Pons, M. (1990). Student differences in self-regulated learning: Relating grade, sex, and giftedness to self-efficacy and strategy use. *Journal of Educational Psychology, 82*, 51–59.

NOTES

Chapter Six

Making Sense of Math

The Process Standards

Social interaction enhances learning.

Learning happens when you can explain your thinking to others in a logical way that they can understand. Individuals must think for themselves to process the information they are learning. Once they have tried to make sense of the material, discussing ideas with others helps. Listening to other people's perspectives enhances learning as well. Sean, the medical school graduate, explained:

At Yale at the medical school, there is no competition. There is no ranking. It is not "this is the top-ranked student." And in the PA program, they would not give us letter grades. It was honors to pass. And if you are anywhere in that range, you pass, and no one is going to know what your GPA is and what your rank is. And it is "we are not ranking you, and you guys are going to collaborate. Because it is not a competition amongst you guys, we want all of you to be great clinicians and great scientists and collaborate with one another," . . . and there is no reason not to teach each other.

You should not be disincentivized from working together; you should be incentivized to work together. That is very interesting because many medical programs have a very competitive outlook. Everyone wants to be the top-ranked student. Because everyone wants the best residency, that is discouraged at Yale, which is really beneficial. There is a sense of collaboration, and I don't think we would have got that if there was competition.

Collaboration is powerful because students can teach each other things and explain things to teach others in meaningful ways. A student did not understand the teacher's explanation; listening to peers who might explain things differently helps.

Collaboration is a powerful method of learning. Students can learn from each other. Liz, the award-winning teacher and parent of STEM graduates, shared:

> I think the elementary classroom is so social, and everything intermingles together. They spend so much time with their classmates that there is so much opportunity to learn from them. In classrooms, there will always be kids who do better in one topic and work on other topics.
>
> I think that if you can learn from your friends, do something. Can it almost be better, right? Because teachers are great, and if these teachers are not teaching in a way that makes sense to someone, but their classroom understands it in a way that they can relate to it, that works better for someone that collaboration is effective. If they learn, . . . that collaboration is so important.

Teachers must consider building communication and peer collaboration in their classrooms.

WHAT CAN TEACHERS DO TO SUPPORT CLASSROOM COMMUNICATION?

The third Common Core Standards for Mathematical Practice (National Governors Association Center for Best Practices, 2010) points out that constructing viable arguments and critiquing the reasoning of others is an important part of learning mathematics. The communication standard is a part of a set of effective math practices that support mathematical learning. All these standards must work together to create the optimal conditions for learning. The Common Core Standards for Mathematical Practice are as follows:

- *Make sense of problems and persevere in solving them.*
  - This means that students solve problems and think about what they are doing instead of blindly following a set of rules through memorization and not understanding why these rules work.
- *Reason abstractly and auantitatively.*
  - Reasoning abstractly and quantitatively involves the ability to represent math as quantities using objects. For example, a student can think about the following problem: Zack has three cookies and gets five more. How many cookies does he have? The quantitative representation is what is represented by the circles. The abstract representation is when the child can use numbers and write a number sentence. Real understanding happens when the child connects the relationship between the circles and the written numerals. A child can reason abstractly when they can use

numbers to represent the three cookies and five and solve the problem: 3 + 5 = 8 (see figure 6.1).

- *Construct viable arguments and critique the reasoning of others.*
  - This practice involves the ability to explain how a math problem was solved and to reflect on another person's idea to analyze whether it makes sense critically.
- *Model with mathematics.*
  - Students should be able to use math in a real-world situation to solve a problem and reflect on the effectiveness of the math model. A model can be a graph, a flow chart, or a formula to solve a problem. For example, when going to the grocery store, students can calculate the price for buying five bananas if each one costs $0.79.
- *Use appropriate tools strategically.*
  - Students should be able to strategically use tools such as pencil and paper, models, rulers, calculators, and spreadsheets.

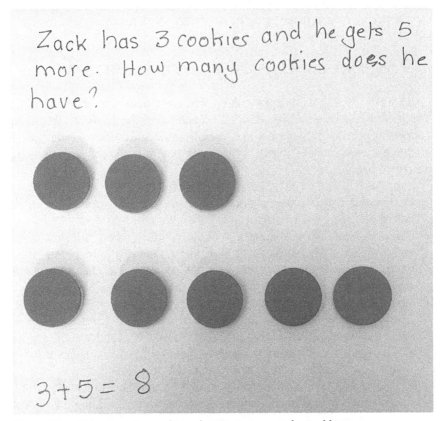

Figure 6.1. Using Counters and Numbers to Represent the Problem
Teruni Lamberg

- *Attend to precision.*
  - Mathematics needs to be explained and communicated precisely so that others can understand. For example, precision involves using accurate terminology such as "place value" and formula, and labeling units of measures or axes on a graph.
- *Look for and make use of structure.*
  - This standard involves looking for patterns or structures—for example, the order you add numbers does not matter. Observing these patterns helps students develop a deeper understanding of larger mathematical concepts, such as the commutative property of addition. For example, adding 7 + 3 is the same as adding 3 +7.
- *Look for and express regularity in repeated reasoning.*
  - Students must look for general methods and shortcuts to solve problems. This means that if they notice something is repeating, such as a decimal when dividing 25 by 8, they can conclude that it must be a repeating decimal. This means that students should pay attention to patterns they notice and try to make sense of what they are doing.

THE WHOLE-CLASS DISCUSSION FRAMEWORK

Teachers can naturally integrate these math practices into their teaching. The book *Work Smarter, Not Harder: A Framework for Math Teaching and Learning* (Lamberg, 2019) outlines specific strategies teachers can use to integrate these practices into their teaching naturally. These strategies are described below.

1. Design of the physical space—organize space for whole-group discussion, individual work, and small-group collaboration.
2. Classroom routines—develop classroom routines where students learn how to listen and communicate their ideas. One of the most important things in developing classroom routines is to build a classroom community where students feel safe to share their thoughts and support each other.
3. Lesson planning—there are three levels of planning to carefully be clear about the big ideas (concepts and skills) students need to learn.
   a. The first level of planning involves identifying what students need to learn and be able to do by examining state standards.
   b. The second level of planning is thinking about how to use class time efficiently to maximize learning opportunities by building on student assessment.

c. The third level of planning involves adapting the lesson to build on student thinking.
4. Whole-class discussion—students should solve problems individually and discuss them in small groups. The teacher can facilitate a discussion to make the big ideas explicit. The focus of the debate should be on a problem or issue that the class is trying to figure out, and the teacher can build on their thinking by introducing new information. First, the teacher might give students a problem to solve. Then the students solve the problem individually, discuss their solutions with their small group, and get feedback. They might revise their thinking as they listen to each other. Once students have worked together, the teacher walks around the room and figures out what students might be confused about and struggling with discussing.

Three Levels of Sense-Making for Discussion

1. Share answers—students share their thoughts.
2. Analyze each other's answers—students analyze each other's responses to see what is the same or different. In doing so, they look at the underlying math concepts and problem-solving strategies.
3. Make the big ideas explicit—the big idea is made explicit by solving the problem. This big idea, such as the associative property, is what the students need to remember to solve additional problems.

CREATING A SUPPORTIVE ENVIRONMENT FOR DISCUSSION

Students need to feel safe and valued when engaging in discussion. Therefore, it is important not to create a classroom environment of competition where students are trying to be better than others. Rather, a spirit of collaboration is helpful. Competition does not build collaboration. Sean, the medical student graduate, explained:

When students compete against each other, they will not help each other, and the power of collaboration is lost. Competition creates the opposite; if I know how to do an integral and figure out a faster way to do it, I will not teach it to someone else because I compete to do that. What is my incentive to help everyone else? It is hard to gain that if the culture is not built for collaboration. If you have a class of 25 and you have competition versus collaboration, you are missing 24 teachers that can help.

Peers are valuable in the learning process if they are collaborative. Students who do not understand the teacher's explanation can get a different perspective from another student.

SUMMARY

Communication is part of the Standards for Mathematical Practice. Collaboration enhances learning through mutual support. When students feel safe sharing ideas and supporting each other, math learning is enhanced.

The Standards for Mathematical Practice (National Governors Association Center for Best Practices 2010) should be a natural way for students to participate in math lessons. Teachers must create a classroom environment

---

Standards for Mathematical Practice

- Make Sense of Problems and Persevere in Solving Them
- Reason Abstractly and Quantitatively
- Construct Viable Arguments and Critique the Reasoning of Others
- Model with Mathematics
- Use Appropriate Tools Strategically
- Attend to Precision
- Look for and Make Use of Structure
- Look for and Express Regularity in Repeated Reasoning

Source: National Governors Association Center for Best Practices, Council of Chief State School Officers. (2010). *Common Core state standards: Math.*

Whole-Class Discussion Framework for Naturally Integrating Math Practices

- Design the physical space so that is is conducive to group collaboration and discussion.
- Develop a classroom routine so that students reflect on their thinking by listening to each other and sharing ideas.

Three Levels of Lesson Planning
- *The First Level of Planning:* Consider the standards that students need to learn (big ideas and how they interrelate).
- *Second Level of Planning:* Plan the lesson that considers assessment builds in collaboration and discussion time.
- *Third Level of Planning:* Adapt lessons based on student learning.

Three Levels of Sense-Making
1. Make thinking explicit: Students share their ideas.
2. Analyze solutions: Students critique each other's thinking and analyze the underlying mathematical concepts.
3. Identify big ideas: Make the math concept explicit so that students can use this concept to solve different problems.

Source: Adapted from Lamberg (2020), *Work Smarter, Not Harder: A Framework for Math Teaching and Learning.* Rowman & Littlefield.

and plan lessons that naturally integrate these math practices. A whole-class discussion framework adapted from Lamberg (2020) outlines an approach for teachers to incorporate these practices naturally. Specifically, this framework contains the following components that teachers can integrate.

What Parents Can Do to Support Their Children

Please encourage students to share their ideas and provide explanations. The goal is to support students to be able to think logically and present an argument. Parents can ask the following questions:

- Why do you think that?
- Can you explain what you did?
- What do you notice?
- Did you check your work?
- Why does it make sense?

BIBLIOGRAPHY

Lamberg, T. (2023). *Sparking the math brain: Insights on what motivates students to learn, creating conditions for learning.* Rowman & Littlefield.

Lamberg, T. (2019). *Work Smarter, not harder: A framework for math teaching and learning.* Rowman & Littlefield.

Lamberg, T., Koyen, L., & Moss, D. (2020). Supporting teachers to use formative assessment for adaptive decision making. *Mathematics Teacher Educator, 8*(2), 37–58.

National Governors Association Center for Best Practices, Council of Chief State School Officers. (2010). *Common core state standards: Math.*

NOTES

Chapter Seven

How Does Learning Happen in the Brain?

Learning happens when the brain makes connections between prior knowledge and new information!

Understanding how the brain learns is beneficial for helping students learn math. Learning happens when the learner makes connections between prior knowledge and new information (see figure 7.1 below). The brain learns and creates memories through new neural pathways. The brain cells connect through a synaptic mechanism through electrical signals (Goswami, 2007). The memory is either stored or forgotten when synaptic connections are made

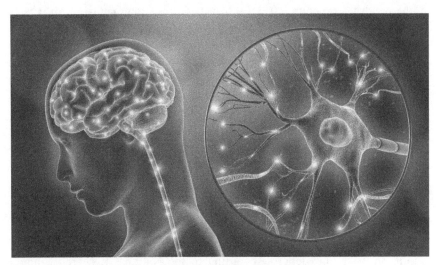

Figure 7.1. Learning Happens When Connections Are Made
Teruni Lamberg

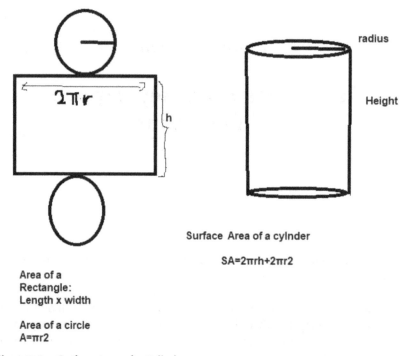

Surface Area of a cylinder

SA=2πrh+2πr2

Area of a
Rectangle:
Length x width

Area of a circle
A=πr2

Figure 7.2.   Surface Area of a Cylinder
Teruni Lamberg

in the dendrites (Klemm, 2013). When there are strong pathways, it creates muscle memory and is stored in the brain (Kennedy, 2016).

When students apply what they know to new situations, learning becomes easier. For example, students can use what they know about finding the area of a circle and a rectangle to figure out the surface area of a cylinder (see figure 7.2). The surface area of a cylinder is made up of two circles and a rectangle. The length of the rectangle is the same as the circle's circumference, as illustrated above. Suppose the student had a deep understanding of finding the surface area of circles and rectangles. In that case, they can certainly transfer that learning to figure out the surface area of a cylinder.

LEARNING WITH UNDERSTANDING

Deep learning represents learning with understanding. Deep learning builds on prior knowledge. According to the National Research Council (Learn, 2000), when deep learning occurs, knowledge can transfer to new situations—for example, understanding how the surface area of two-dimensional objects relates to finding the surface area of three-dimensional objects. Learning the area of a cylinder is meaningful if it builds on the previous understanding of finding the area of a circle and rectangle. If this prior knowledge is lacking, then memorizing the formula will not make sense.

According to the National Research Council (Learn, 2000), you can't transfer learning to new situations if your initial learning was weak. For example, if a student had memorized the formula for the area of a cylinder and rectangle and did not understand why the formula works, they would not be able to use that knowledge to figure out the surface area of a rectangular prism. They won't be able to transfer their learning to a new situation.

TEACHING TO THE TEST IS NOT HELPFUL IF IT DOES NOT BUILD UNDERSTANDING

Many parents and teachers feel pressured to help their children perform well on tests. Therefore, some teachers and parents focus on students learning test-taking skills to perform well on standardized tests. The reality is that teachers and schools are measured by how well students perform on standardized tests. One of the most ineffective ways to support learning is to solely focus on test-taking skills and ignore conceptual understanding and developing higher-level thinking skills.

When a child feels pressured to perform well on tests without focusing on understanding what they are doing, this type of behavior leads to math anxiety and even failure. Furthermore, focusing only on test-taking skills through rote memorization is ineffective because memorizing random things without connecting them to anything familiar only focuses on the short-term memory, making recall more difficult.

Students need to understand "Why does it work? How does it make sense?" When things make sense, it is easier to remember. Consider the two images in figure 7.3. Which picture is easier to remember? The tic-tac-toe board or the scrambled-up tic-tac-toe board?

Remembering the tic-tac-toe game is much easier because everyone can recognize the pattern. When the board is cut up and jumbled, it does not make sense or connect to anything you know unless you are told it is a cut-up tic-tac-toe board. So, remembering the random pattern takes much more effort than remembering something familiar. It would be easier to reproduce if a person started noticing the pattern.

Which is easier to remember?

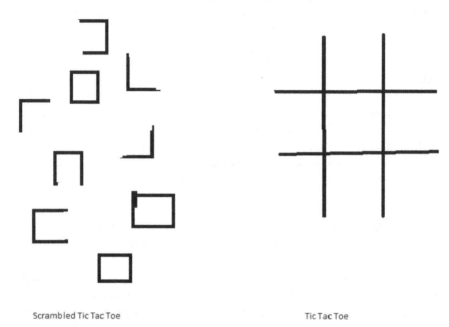

Scrambled Tic Tac Toe   Tic Tac Toe

Figure 7.3.   Learning Through Pattern Recognition
Teruni Lamberg

METACOGNITION

Learning should be active. Active learning has proved to be more effective than being a passive learner (Tanner, 2012). When students reflect on what they learned and struggled with, it empowers them to act. In other words, students should reflect on what they know and figure out what they struggle with, and their learning style empowers them to take an active role.

Students can try to solve problems on their own and explore solutions. According to Dewey (1933), thinking is triggered by situations that evoke perplexity, confusion, or doubt. Therefore, students must engage in problem-solving and reflect on their thinking solutions.

Parents and teachers can encourage students to solve problems and reflect on their thinking. What were they doing and why? Which parts were difficult or easy? What do they still need to do to solve the problem?

The bottom line is that every child can learn math when the lessons are built on the student's prior knowledge and support higher-level thinking. For example, if a student understands how to multiply by one-digit numbers, it would be easier to help them learn how to multiply with two-digit numbers.

If the student has a grasp on how to multiply by one-digit numbers such as 5 × 6, solving problems such as 26 × 4 would be difficult and harder to learn. Teachers must consider students' prior knowledge and build a bridge to more sophisticated thinking so that the students see connections between what they know and new information.

- There is a standardized curriculum and there are standardized standards, but there are not standardized kids. True learning builds a bridge between the student's prior knowledge and new information. When new understanding does not connect to prior knowledge, it does not stick.
- Teachers should encourage students to engage in problem-solving by allowing them the flexibility to think about the problem in ways that make sense to them.

Students should be encouraged to think through the problem and solve it using what they know in a way that makes sense to them. For example, a student might solve 78 + 43 by making friendly numbers and thinking about the problem as 80 + 41. Another student might approach it by adding 70 + 40 and then adding 8 + 3. Another student might use place value blocks to solve the problem. Encouraging students to think throughout the problem builds the critical thinking skills needed to make sense of the math. They need to ask *why* it works. When math makes sense, it leads to deeper learning. This approach contrasts with the teacher showing students the steps to solve the problem and expecting the problems to be solved only using the teacher's strategies regardless of whether it makes sense to the student.

PRACTICE MAKES PERFECT!

Practice is an important part of learning. Without practice, it is hard to remember information and learn the material. One of the mathematicians pointed out that students should study and practice problems daily to develop a deep understanding and learn the material. He shared:

> And so, you can learn something today, and it will be gone by tomorrow unless you practice and practice. And then you develop a brain muscle; you create a place in your brain to store that information.

Similarly, Greta the scientist shared that those kids who participate in organized sports or music are much more familiar with the concept of practice. That repetition is needed to get good at something. Students must learn that it is important to do the homework to learn.

Actually, if you want to be on a good path, you need to practice! You can watch the person play the piano a thousand times, and you will still not know how to play the piano. You are only going to learn to play the piano when you consistently practice playing the piano. The same thing is true with math, chemistry, and science.

Students need to know that practice and repetition are part of becoming good at something and developing math skills.

EMOTION AND LEARNING

Emotion plays a role in learning. Students are more likely to learn when they are excited about and interested in their learning and see the relevance. On the other hand, if the student goes to class and does not want to be there and is afraid of math or feels like the teacher does not like them, it is much harder to put forth the effort to learn.

Mary Helen Immordino-Yang and Antonio Damasio (2007) wrote a book chapter on the role that emotions play in thinking. They point out that most people think emotion is separate from learning, while they are interrelated. Emotions guide what people do and how they think. The messages students hear about their capabilities and learning math impact and shape how they feel. How students feel impacts their learning and the effort they put forth. Immordino-Yang and Damasio also point out that more research is needed in this area.

SUMMARY: HOW DOES LEARNING HAPPEN IN THE BRAIN?

- Learning happens when the individual connects their prior knowledge and new information.
- Information is retained when it makes sense and connects to prior knowledge.
- Practice helps with memory—students get better with practice.
- Emotion plays a role in learning. If an individual is excited about what they are learning, they are more likely to persist. Therefore, math learning should become an enjoyable experience, like solving a puzzle. Students need to believe in themselves.

BIBLIOGRAPHY

Dewey J. (1933). *How we think: A restatement of the relation of reflective thinking to the educative process.* Heath.

Goswami, U. (2007). Neuroscience and education. In Fischer & Immordino-Yang, *The Jossey-Bass reader on the brain and learning.* John Wiley & Sons.

Immordino-Yang, M. H., & Damasio, A. (2007). We feel, therefore we learn: The relevance of affective and social neuroscience to education. *Mind, Brain, and Education, 1*(1), 3–10.

Kennedy, M. B. (2016). Synaptic signaling in learning and memory. *Cold Spring Harbor perspectives in biology, 8*(2), a016824.

Klemm, W. R. (2013). *Core ideas in neuroscience.* Smashwords.

Learn, H. P. (2000). *Brain, mind, experience, and school.* Committee on Developments in the Science of Learning.

Tanner, K. (2012). Promoting student metacognition. *CBE Life Sciences Education, 11*(2), 113–120. https://doi.org/10.1187/cbe.12-03-0033

Zohar (2009). Paving a clear path in a thick forest: A conceptual analysis of a metacognitive component. *Metacognition Learning, 4,* 177–195.

NOTES

Part II

PROGRESSION OF MATHEMATICS STANDARDS, K–5

Chapter Eight

Number and Operations in Base 10

This section outlines the progression of math concepts that students learn across the grade levels as identified in the Common Core Mathematics Standards (CCSM, 2010). Understanding what students should learn at each grade level is helpful for parents and teachers to support students. Parents and teachers can identify what gaps students might have, and what they can do to challenge students.

Strategies for looking at the standards include:

- Identify standards for each grade level
- Figure out what the students need to understand mathematically
- Examine assessments to figure out how to challenge or support students with gaps
- Examine suggested resources for deeper exploration of the standards

Students should develop proficiency in adding, subtracting, multiplying, and dividing across grade levels. Table 8.1 shows the progression students follow to learn across grade levels. Each grade level sets the foundation for more sophisticated learning. Learning to add, subtract, multiply, and divide involves not only learning a set of rules; students also need to understand the underlying concepts of addition (see tables 8.2 and 8.3 and figure 8.1). They are supposed to know the formal rules (algorithm) for solving math problems when they reach fourth grade. Each grade level sets the foundation for progressively developing a conceptual understanding of math.

Table 8.1.   Progression of Learning Addition across Grade Levels

Grade	Expected Proficiency at Each Grade Level
K	Add and subtract within 5
1	Add and subtract within 10
2	Add and subtract within 20 Add and subtract within 100
3	Multiply and divide within 100
4	Add and subtract within 1,000 (using a standard algorithm)
5	Multiply multidigit numbers (using a standard algorithm) Learn decimals

Table 8.2.   Understanding Counting, Addition, and Subtraction

Addition	Modeling Strategies	Example
Counting	Using physical tools such as blocks, counting with fingers, drawing, tally marks, ten frames, and 100 charts • Count all • Count on from first • Count on from larger	7 + 8 = ? Count objects individually, "One, two, three . . . ," while pointing. Make a pile with 7 and another with 8 and count them all—together. Count the first number in the problem and then add the next. 7, count out 8 blocks, start at 7 and go 8, 9 . . . until 15. Start at 8 and then add on 7.
	Understanding place value and digits and their role in addition	The big idea is understanding that the whole is made up of parts. If you combine the parts, it makes up the whole. 20 + 38 58
	Understanding how to regroup	Requires an understanding of place value.

Table 8.3. Progression of Subtraction

Subtraction	Equation	Progression
Modeling with blocks, figures, and drawings	7 − 5 = ?	Understand that subtraction involves taking a part away from the whole. You need to understand what the whole is and what the part represents. 7 represents the whole, and 5 represents the part that is taken away.
Using numbers	27 − 8 = ?	Compose and decompose numbers to make friendlier numbers. For example, think about the problem 27 − 8 as 20 + 7 and the 8 is made up of 7 + 1. Use place value and regrouping. For example, 27 − 7 = 20 and subtract 1 more to get 19. Standard algorithm using regrouping.

Place Value		Connect model and written Numbers			
Count by ones	Count by ones	●●●●●●● ● 7 (count in sequence 1, 2,3 and know that 7 is a set of seven objects)			
Base 10	Count by 10's Count by 100.' Count by 1000.'	1000's Place	100's place	10's Place	1's
		1	3	1	2
		1312 is made up of 1000 +300 +10 + 2			
Decimals	Decimal point Place Value 1/10 Place Value 1/100 Place value 1/1000	Decimal Point	1/10 Place	1/100's place	1/1000 Place
		.	.01	.001	.001

Figure 8.1. Place Value: Understanding What Value the Digit Represents
Teruni Lamberg

STRATEGIES FOR LEARNING MULTIPLICATION

Helping students learn their multiplication tables involves building understanding and fluency. Start with counting strategies and encourage students to use logical reasoning to figure out facts they do not know (see table 8.4 and figure 8.2). When students develop reasoning strategies and understand how multiplication works, through practice they will become more efficient and develop speed. Focusing on speed and timed tests only stresses students and does not help with recall (Van de Walle et al., 2018).

Table 8.4.   Strategies for Multiplication

Multiply	Strategies (Different Ways to Solve Problems)	Example
0	Any number multiplied by zero will be zero.	5 × 0 = 0
1	Any number multiplied by one will remain the same.	26 × 1 =26
2	Double the number (add it twice).	6 + 6 = 6 × 2 6 × 2 =12
3	Double the number and add one more set of numbers.	3 × 5 = ? 5 × 2 = 10 (double 5) 10 + 5 = 15 (add 5)
4	Double the number and double it again.	8 × 4 = (8 × 2) + (8 × 2) 16 + 16 = 32
5	Multiply by 10 and divide by half.	5 × 5 =? 5 × 10 = 50 Half of 50 is 25
6	Multiply the number by three and double it.	6 × 4 = 24 (3 × 4) = 12 double is 24
7	Break seven up into friendlier numbers using the distributive property.	7 × 6 = 42 7 × 5 = 35 7 × 1 = 7 (7 × 5) + (7 × 1) = 42
8	Multiply a number by four and double it.	8 × 4 = 32 You can think about 4 × 4 = 16. If you double 16, you get 32.
9	The digits in the 9s in the multiplication table add up to 9. The answer in the 10s digit is one less than the number you are multiplying.	9 × 7 = 63 Look at the multiplication table for 9s. Notice the digit in the 10s place is one less than 7 for 9 × 7. It is 6 in the 10s place. This means the digits in the 10s place and 1s place need to add up to 9. The digits 6 + 3 = 9. The answer is 63.
10	If you multiply any number by 10, add a zero.	7 × 10 = 70

	Big ideas
Connect addition and multiplication ❖ Build models/arrays ❖ Connect to written symbols (e.g., 2 × 3 = 6) ❖ Notice patterns in a 100's chart	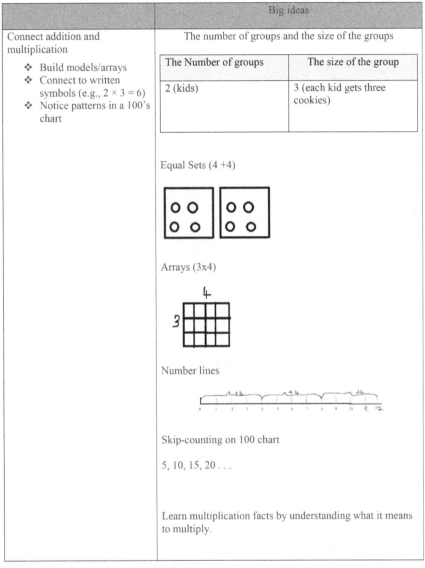

Figure 8.2.   **Progression of Multiplication**

	Multi-digit numbers Informal Method · Standard Algorithm 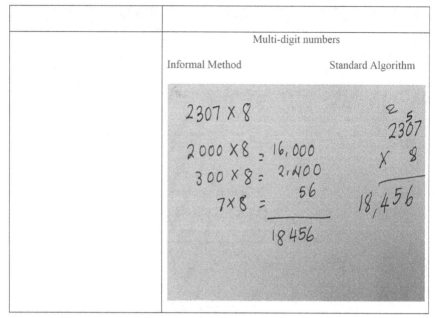

Figure 8.2. Progression of Multiplication

RESOURCES

Websites (Virtual Manipulatives and Games)

- National Library of Virtual Manipulatives: http://nlvm.usu.edu/en/nav/vlibrary.html
- Math and Logic Problems: https://aplusclick.org/
- Math Playground: www.mathplayground.com/logic-games.html
- Math Is Fun: www.mathsisfun.com/numbers/index.html
- IXL: www.ixl.com/math

YouTube Videos

Many websites contain YouTube videos. Type "addition and subtraction" or "multiplication and division" on YouTube to find videos. The following are some examples:

- Numberblocks—Subtracting with the Numberblocks: www.youtube.com/watch?v=mcl1uzbn3QQ&t=259s
- Numberblocks—#Back to School, Meet Numbers 11–15: www.youtube.com/watch?v=1QO5AklawjQ

Books

Carpenter, T. P., Fennema, E., Franke, M. L., Levi, L., & Empson, S. B. (2015). *Children's mathematics: Cognitively guided instruction*. Heinemann.
Number talks: *Whole number computation*. Heinemann.
Parrish, S. (2011). *Number talks*. Math Solutions.

BIBLIOGRAPHY

National Governors Association Center for Best Practices, Council of Chief State School Officers. (2010). *Common Core State Standards: Math*.
van de Walle, J. A., Karp, K. S., & Bay-Williams, J. M. (2018). *Elementary and middle school mathematics: Teaching developmentally* (10th ed.). Pearson.

NOTES

Chapter Nine

Algebraic Thinking

Algebraic thinking involves looking for patterns and making generalizations. Students can discover rules by solving problems and noticing if the rule is always true. For example, a teacher can ask "What happens to a number when you add zero? Would it be the same every time?" By noticing patterns, students can make sense of the rules. For example, a student could notice that 25 + 26 is made up of 25 + 25 + 1.

Algebraic thinking involves seeing the relationship between numbers. Students could use what they know about adding 25 + 25 and one more. Another big idea is the concept of equality. The equal sign means "same as," instead of using a symbol before the answer. Students also explore the concept of inequality by learning about relationships that are "greater than or less than."

In the lower grades students are informally introduced to variables, and then they learn to use symbols to represent variables in later grades. The following section describes some rules students can explore and discover when learning math.

PROBLEM TYPES

Note that the problem types in tables 9.1–9.3 increase in complexity and lay the foundation for algebraic thinking. Figure 9.1 depicts the progression of numbers and operations through the elementary grades.

Table 9.1. Addition and Subtraction Problem Types

Problem Structure	Number Sentence	Word Problem
a + b = _	3 + 4 = __	Zack has three cookies, and he gets four more. How many cookies does he have altogether?
_ + b = c	_ + 4 = 7	Zack has some cookies, and he gets four more. Now he has seven cookies. How many cookies did he have?
a + _ = c	3 + _ = 7	Zack has three cookies, and he gets more cookies. How many cookies did he get?
a – b = ?	8 – 5 = ?	Zack had eight cookies and he ate five. How many does he have left?
_ – b = c	_ – 5 = 3	Zack had some cookies, and he ate four. Now he has three left. How many cookies did he have?
a + _ = c	8 -=– _ = 3	Zack has eight cookies. He ate some cookies. Now he has three left. How many cookies did he eat?

Table 9.2. Multiplication Problem Types

	Unknown Product	Group Size Unknown	Number of Groups Unknown
Equation	3 × 5 = 15 a × b = ?	3 × ? = 15 and 15 ÷ 3 = ? a × ? = b and b ÷ a = ?	_ × 5 = 15 and 15 ÷ 3 = ?
Equal groups	There are 3 baskets with 5 apples in each. How many apples are there altogether?	If 15 apples are shared equally and packed inside 3 baskets, how many apples will be in each bag?	How many bags are needed if 15 apples are to be packed with 3 in each bag?
Arrays	There are 3 rows of apples with 5 in each row. How many apples are there?	If 15 apples are arranged into 3 equal rows, how many apples will be in each row?	How many will there be if 15 apples are arranged into equal rows of 5?
Compare	A baseball costs $5. A soccer ball costs 3 times as much as a baseball. How much does the soccer ball cost?	A soccer ball costs $15, which is 3 times as much as a baseball. How much did the baseball cost?	A soccer ball costs $15, and a baseball costs $5. How many times more does the soccer ball cost than the baseball?

Table 9.3. Measurement Division and Partitive Division

Measurement Division	Partitive Division
The whole amount and the number of groups are known.	The whole amount and the size of the group are known.
• What is the size of the group?	• What is the number of groups?
There is a 16-inch piece of ribbon. How many 4-inch pieces can you cut?	There are 16 pieces of candy. How many pieces would each child get if 4 children shared it equally?

Grade	Standard	Examples
K	Counting (Names, sequence, Cardinality (how many in a set of objects)	Check out the following video (subitize Rock) by Jack Hartmann https://www.youtube.com/watch?v=ib5Gf3GIzAg
	Compare numbers (bigger, smaller)	Check out Comparing Numbers for Kids—Greater than Less than https://www.youtube.com/watch?v=E34PAOGYRNk
	Understand addition and subtraction—combine things and take them apart.	$4 + 2 = 6 \quad 6 - 4 = 2$
1	• Represent and solve problems using addition and subtraction within 20 • Add three numbers that add less than or equal to 20. • Understand what it means to add • The equal sign represents a relationship between numbers	See problem types on the previous page. $4 + 3 + 7 = 14$ $3 + 2 = 4 + 1$

Figure 9.1.   Progression of Numbers and Operations
Teruni Lamberg

2	Solve addition and subtraction two-step problems.	▶ *Results Unknown*
		  ▶ There are 29 gumballs in the jar. Eighteen more are added. How many are there now?
		  ▶ *Change Unknown*
		  ▶ There are 29 gumballs in the jar. Some more gumballs are added. Now there are 47 gumballs. How many gumballs were added to the jar?
		▶ *Start Unknown*
		  ▶ There are some gumballs in the jar. Eighteen more gumballs were added to the jar. There are now 47 gumballs. How many gumballs were already in the jar?

	Equal groups	Skip count on 100 charts by numbers such as 5, 10, 15, etc. Understand what it means to count equal sets (laying a foundation for multiplication).
	Arrays	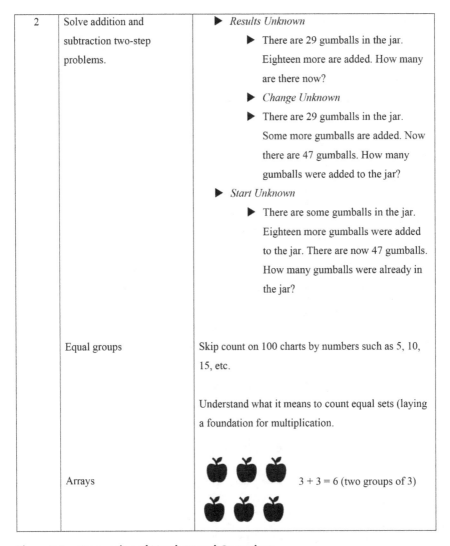 3 + 3 = 6 (two groups of 3)

Figure 9.1. Progression of Numbers and Operations
Teruni Lamberg

3	Fluently multiply and divide within 100. Learn multiplication facts.	(See problem types for multiplication)
	Understand the relationship between multiplication and division.	Understand 3 × 5 = 15 15 ÷ 3 = 5 3 × 5 (Size of group multiplied by the number of groups) 15 ÷ 3 = 5 (the whole split equally into three equal groups. The answer is the size of the group.)
4	Use addition, subtraction, multiplication, and division to solve problems.	Multistep word problems with remainders
	Multiplication as a comparison.	Multiplicative thinking e.g., interpret 30 = 5 × 6 "thirty is five times as many as six and six times as many as five" The whole number is a multiple of each factor
	Factors and multiples	E.g., 21 = 7 × 3
	Generate and analyze patterns	Given a rule, identify partners in a sequence.
5	Write and interpret numerical expressions. Analyze patterns and relationships	Use parenthesis, brackets, or braces 4 × 5 + 7 (why the order of operations matters) Write simple expressions

Figure 9.1. Progression of Numbers and Operations
Teruni Lamberg

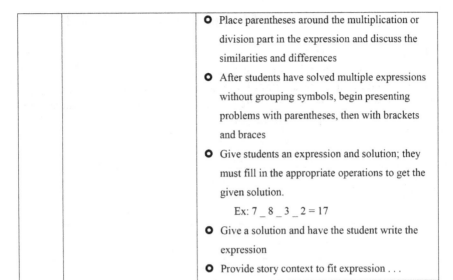

		O Place parentheses around the multiplication or division part in the expression and discuss the similarities and differences
		O After students have solved multiple expressions without grouping symbols, begin presenting problems with parentheses, then with brackets and braces
		O Give students an expression and solution; they must fill in the appropriate operations to get the given solution. Ex: $7 \_ 8 \_ 3 \_ 2 = 17$
		O Give a solution and have the student write the expression
		O Provide story context to fit expression . . .

Figure 9.1. Progression of Numbers and Operations
Teruni Lamberg

UNDERSTANDING OPERATIONS AND PROPERTIES

The key to algebraic thinking is not simply memorizing definitions but exploring patterns by trying out examples and noticing partners. For example, a child can try adding zeros to multiple numbers and decide whether they get the same answer every time. Table 9.4 lists the properties of addition, and table 9.5 lists the properties of multiplication.

Table 9.4. Properties of Addition

Property of Operation	Equation
Commutative property	$a + b = b + a$ Example: $3 + 2 = 2 + 3$
Associative properties	$(a + b) + c = a + (b + c)$ Example: $(3 + 2) + 5 = 3 + (2 + 5)$
Additive identity	$a + 0 = 0 + a = a$ Example: $5 + 0 = 0 + 5 = 5$
Additive inverse	$a - a = a + (-a) = 0$

Table 9.5. Properties of Multiplication

Property of Operation	Equation
Commutative property of multiplication	$a \times b = b y \ a$ Example: $3 \times 2 = 2 \times 3$
Associative properties of multiplication	$(a \times b) + c = a + (b \times c)$ Example: $(3 \times 2) + 5 = 3 \times (2 \times 5)$
Inverse	$a \times 1/a = 1/x \times a = 1$
Distributive property for multiplication	$a \times (b = c) = a \times b + a \times c$

ONLINE RESOURCES AND VIRTUAL MANIPULATIVES

Didax.com Virtual Manipulatives: www.didax.com/apps/ten-frame
Annenberg Learner: www.learner.org/series/learning-math-patterns-functions-and -algebra/algebraic-thinking
Khan Academy: www.khanacademy.org/kmap/operations-and-algebraic-thinking-j
Illustrative Mathematics, n.d. *Grade 4—Operations and Algebraic Thinking*: http:// tasks.illustrativemathematics.org/content-standards/4/OA
Illustrative Mathematics, n.d. *Grade 4—Rounding to the Nearest 100 and 1000*: http://tasks.illustrativemathematics.org/content-standards/4/NBT/A/3/tasks/1806

BIBLIOGRAPHY

Blanton, M. (2008). *Algebra and the elementary classroom: transforming thinking, transforming practice*. Heinemann.

Blanton, M., Zbiek, R. M., Dougherty, B., Crites, T., & Levi, L. (2011). *Developing essential understanding of algebraic thinking for teaching mathematics in grades 3–5. Series in essential understandings*. National Council of Teachers of Mathematics.

Carpenter, T. P. (2012). *Thinking mathematically: Integrating arithmetic and algebra in elementary school*. Heinemann.

National Governors Association Center for Best Practices, Council of Chief State School Officers. (2010). *Common core state standards: Math*.

Small, M. (2014). *Uncomplicating algebra to meet common core standards in math, K–8*. Teachers College Press, Nelson Education.

van de Walle, J. A., Karp, K. S., & Bay-Williams, J. M. (2018). *Elementary and middle school mathematics: Teaching developmentally* (10th ed.). Pearson.

NOTES

Chapter Ten

Number and Operations: Fractions

The early grades (K–2) lay the foundation for learning fractions. Students learn to partition geometric figures into equal areas such as half and fourths. Students learn about the concept of equal-sized groups and the meaning of part/whole when learning to add and subtract. In third grade students formally learn about fractions as part of a whole. In fourth grade, students learn to add and subtract fractions and work with common and uncommon denominators. In fifth grade students learn about fractions in fair-sharing situations, which involves dividing a whole into equal parts. Figure 10.1 depicts this progression through elementary grades.

BIG IDEAS RELATED TO UNDERSTANDING FRACTIONS

The following concepts are important for learning fractions and cut across grade levels:

- *Unit:* How the whole is visualized in a problem context influences the answer. "What is the whole?" It could be one pizza, two pizzas, or a dozen cookies.
- *Partition* is the process of sorting or cutting a whole unit into equal parts to solve a problem. "How do you cut the whole unit into equal parts?" or "How do you sort the whole unit into equal parts?"
- *Equivalency*: The whole has been split into equal-sized pieces:

Grade	Concept	Example
3	Understand fraction as representing a part of a whole	For example, the fraction 1/4 represents 1 out of four equal parts that make up the whole. NUMERATOR: represents the number of parts of a whole. DENOMINATOR is the bottom number of the fraction. It represents how many equal parts the whole is divided into. **Another example:** The whole is divided into four equal parts. One of the parts represents the fraction ¼. The one refers to the numerator, and the 4 represents the number of equal parts the whole is divided into. It represents the denominator.
4	Add and subtract fractions Equivalent fractions	Draw models or use fraction manipulatives to understand the concept of adding make connections to the number sentence. 1/3 +1/3 =2/3

Figure 10.1. Progression of Fraction Standards Across Elementary Grades
Teruni Lamberg

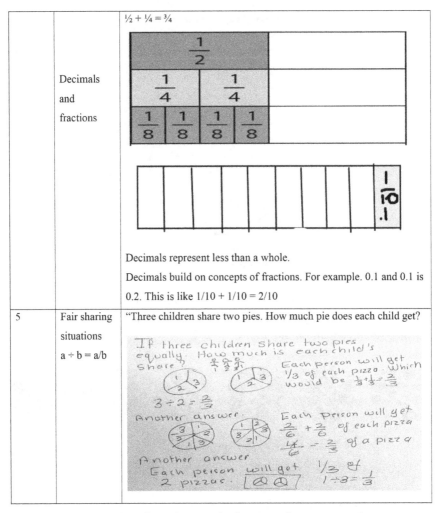

Decimals and fractions	$\frac{1}{2} + \frac{1}{4} = \frac{3}{4}$

Decimals represent less than a whole.

Decimals build on concepts of fractions. For example. 0.1 and 0.1 is 0.2. This is like $1/10 + 1/10 = 2/10$

| 5 | Fair sharing situations $a \div b = a/b$ | "Three children share two pies. How much pie does each child get? |

Figure 10.1. Progression of Fraction Standards Across Elementary Grades
Teruni Lamberg

FRACTION MODELS

- *Area Model:* This model helps represent a whole unit that needs to be split into equal pieces (drawings, manipulatives). See figure 10.2.

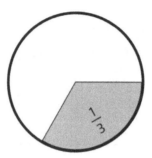

Figure 10.2.   Area Model
Teruni Lamberg

- *Set Model:* This model helps represent discrete units where the whole is sorted, such as six pieces of candy. A picture of the object can represent the problem situation. See figure 10.3.

Figure 10.3.   Set Model
Teruni Lamberg

- *Linear Model:* A bar or a number line can be used to represent length or distance, such as running a race or a plank of wood that needs to be cut. See figure 10.4.

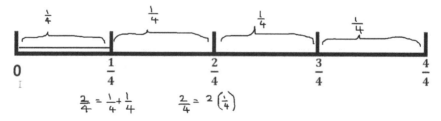

Figure 10.4.   Number Line
Teruni Lamberg

Key ideas for understanding number lines include the following:

• A fraction is a number on the number line, like a whole number. It represents the distance from zero.
• The denominator of a fraction represents how many equal pieces the whole was split into.
• Realize that the number line extends both ways.

RESOURCES

Virtual Manipulatives

Percentage Strips: https://toytheater.com/percentage-strips/
Fraction Bars | Math Playground: https//mathplayground.com
National Library of Virtual Manipulatives: http://nlvm.usu.edu/en/nav/vlibrary.html
Fraction Game—Illuminations (National Council of Teachers of Mathematics): www.nctm.org/Classroom-Resources/Illuminations/Interactives/Fraction-Game/
Fraction Models: www.nctm.org/Classroom-Resources/Illuminations/Interactives/Fraction-Models/
Illustrative Mathematics, n.d. *Grade 3—Locating Fractions Less Than One on the Number Line*: http://tasks.illustrativemathematics.org/content-standards/3/NF/A/2/tasks/168
Illustrative Mathematics, n.d. *Grade 3—Find ¼ Starting from 1, Assessment Version*: http://tasks.illustrativemathematics.org/content-standards/3/NF/A/2/tasks/1350

Videos

Khan Academy, Understand Fractions: www.khanacademy.org/math/arithmetic/fraction-arithmetic

Online Resource Explaining Common Core Standards

Mathematics Framework. In *2013 Mathematics Framework Chapters*. California Department of Education. www.cde.ca.gov/ci/ma/cf/mathfwchapters.asp

BIBLIOGRAPHY

California Department of Education. (2013). 2013 Mathematics Framework chapters. www.cde.ca.gov/ci/ma/cf/mathfwchapters.asp

National Governors Association Center for Best Practices, Council of Chief State School Officers. (2010). *Common core state standards: Math.*

Parrish, S. (2022). *Number talks: Fractions, decimals, and percentages.* Heinemann.

Small, M. (2014). *Uncomplicating fractions to meet common core standards in math, K–7.* Teachers College Press.

van de Walle, J. A., Karp, K. S., & Bay-Williams, J. M. (2018). *Elementary and middle school mathematics: Teaching developmentally* (10th ed.). Pearson.

Chapter Eleven

Measurement and Data Analysis

The measurement and data analysis standards focus on measurement, where students learn about units and measure objects. Initially, they work with nonstandard units and progress to learning more sophisticated standard units of measure across the grades. The nonstandard units involve iterating a unit multiple times to make a measurement and knowing where to place the items to measure. They progress to using more sophisticated tools. This standard also includes learning how to tell time and about money. (See figure 11.1.)

Students learn how to represent and interpret data in categories to answer a question, connect measurement to collecting data, and connect data to addition and subtraction.

Data analysis involves looking at trends and patterns in data to answer a question. It is more than calculating the mean or creating a graph. It is about collecting, organizing, and interpreting data to answer questions to make decisions. The following section describes what students need to understand about the process of data analysis.

Students need to understand the rationale as to why people collect and analyze data. The section below outlines the process involved in collecting and analyzing data. This process cuts across grade levels.

1. Formulate a good question to address a problem or issue. Consider the audience.
2. A data creation discussion should take place. The student must decide on what data to collect to address the problem or issue. For example, how should the data be collected to answer the question "How much TV did you watch last week?" Is it consistent? Do we include the weekends? Does having the TV on while doing chores count? Come up with a plan to collect data.

GRADE		Example
K	Describe and compare measurable attributes of objects such as length and weight. Compare two objects with measurable attributes	"Which is longer?" (Compare three lengths 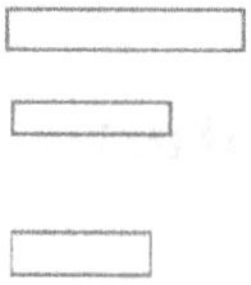 "Which animal is heavier? 
1	Use standard units to *measure lengths*. Connect addition and subtraction to measuring length. Time: Tell and write the time in hours and half hours in analog and digital clocks. Organize, represent, and interpret data	 • When measuring needs to measure from end to end without gaps • Understand what a standards unit is, such as feet For example, "three o'clock, three thirty" in analog and digital clocks Create pictures and bar graphs representing data (up to four categories).
2	Standard Units—measure length using tools Estimate length (inches, feet, centimeters, and meters) Measure and compare objects of different lengths and compare the difference.	Tools such as rulers, yardsticks, meter sticks, and measuring tapes. Relate addition and subtraction to length "Which piece of yarn is longer?" "How much longer is the piece of year?"

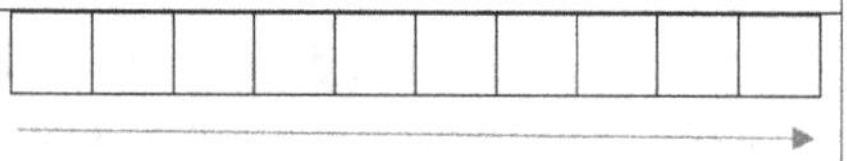

Figure 11.1. Progression of Measurement and Data Analysis
Teruni Lamberg

	Time and money	Write time in analog and digital clock for the nearest 5 minutes
	Represent and interpret data • Picture graph • Bar Graph o Line Plot	Create a line plot of an object measured repeatedly to track changes such as a plant's growth.
3	Time	Tell and write time to the nearest minute and measure time intervals. Solve time interval problems involving addition.
	Represent and interpret data	Draw scaled pictures and bar graphs with several categories. Also, use scale bar graphs such as each bar can represent five pets.
	Area (connect to arrays) perimeter	Understand the difference between area and perimeter. Find the area of a rectangle. The area of the rectangle is 4 x 3=12 It can also be (2 x 3) x 2 You can have the same perimeter but different areas.
4	Measurement problems (conversion)	Convert Feet to inches and inches to feet
	Solve problems involving distance, intervals of time,	Use diagrams • Number lines that feature a

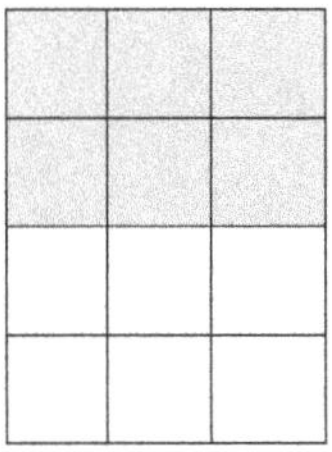

Figure 11.1. Progression of Measurement and Data Analysis
Teruni Lamberg

		liquid volume, masses of objects, and money (use four operations, fractions, and decimals)	measurement scale
		Use formulas for the rectangle in real-world problems • Area = Length × Width • Perimeter = (L × W) × 2 Represent and Interpret data • Line plot	Make a line plot with fraction measures (1/2, 1/4, 1/8) Formed with two rays with a common endpoint.
		Angles	An angle of 1/360 of a circle represents one degree (of a circle). The angles could be added up to make a larger angle or split into smaller angles (additive).
		• Use protractor	Measure using a protractor
5		Convert measurements	Centimeters to meters (e.g., 5 cm is 0.05 meter) Use conversions to solve a multistep word problem
		Line plot	Line plot with fractions of unit (1/2, 1/4, 1/8) Use fraction operations to problem solve using the information on the line plot. For example, if two identical beakers contain unequal amounts of liquid, how much liquid will each beaker contain if the liquid is divided equally?
		Volume	Understand the concept of volume using unit cubes. Connect formula to model V = l × wlh V = b × h for rectangular prisms

Figure 11.1.   Progression of Measurement and Data Analysis
Teruni Lamberg

3. Analyze data. Organize data to make sense of it. Students can analyze the pros and cons of different forms of representing data such as a line plot or a bar graph.
4. Interpret and analyze the organized data to answer the question in step 1.

Across grade levels, students learn how to represent data in more sophisticated ways and understand how to interpret the data and the conventions of creating graphs and charts.

TYPES OF DATA

• Numerical data: continuous data (e.g., the temperature, which goes up and down)
• Categorical data: groups by categories with labels (e.g., favorite season).

TYPES OF GRAPHS

Figures 11.2–11.4 show examples of a picture graph, a bar graph, and a line plot.

Figure 11.2.   Picture Graph
Teruni Lamberg

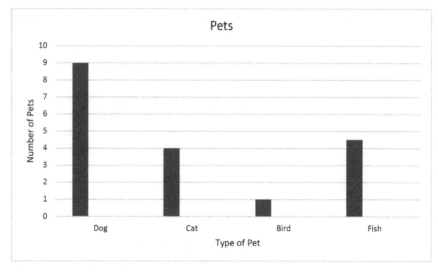

Figure 11.3. **Bar Graph**
Teruni Lamberg

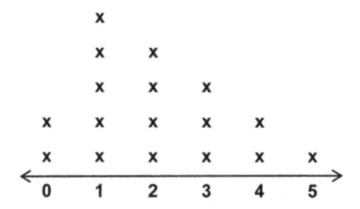

Figure 11.4. **Line Plot**
Teruni Lamberg

ONLINE RESOURCES

IXL Math: www.ixl.com
Desmos: www.desmos.com
Toy Theater—Clock: https://toytheater.com/clock
Illustrative Mathematics, n.d. Comparing Money Raised: http://tasks.illustrative
mathematics.org/content-standards/4/OA/A/2/tasks/263.

Online Resource Explaining Common Core Standards

Mathematics Framework Chapters. In *2013 Mathematics Framework Chapters*. California Department of Education. www.cde.ca.gov/ci/ma/cf/mathfwchapters

BIBLIOGRAPHY

Bright, G. W., & Friel, S. N. (2004). *Navigating through data analysis*. National Council of Teachers of Mathematics.
National Governors Association Center for Best Practices, Council of Chief State School Officers. (2010). *Common core state standards: Math*.
Van de Walle, J. A., Karp, K. S., & Bay-Williams, J. M. (2018). *Elementary and middle school mathematics: Teaching developmentally* (10th ed.). Pearson.

NOTES

Chapter Twelve

Geometry

Students in early grades reason about two-dimensional and three-dimensional shapes. As they progress across grade levels, they reason about the attributes of shapes. In third and fourth grade, they start learning about angles. In fifth grade, they progress to learning about location and the coordinate plane. Figure 12.1 shows this progression of geometry standards. Farther down, figure 12.2 offers a description of shapes and their attributes, and figure 12.3 depicts lines and angles as third and fourth graders would learn about them.

Grade	Standard	Example
K	Identify and describe squares, circles, triangles, rectangles, hexagons, cubes, cones, cylinders, and spheres. Know the difference between two- and three-dimensional shapes. Analyze, compare, create, and compose shapes.	"What shapes do you see in our room?" Where are these shapes located? Identify shapes in the physical environment and use terms such as above, below, in front of, behind, and next to. Describe shapes in the environment according to their attributes. Such as, a square has four sides. Build shapes using toothpicks and marshmallows to represent three-dimensional objects. Look for examples and non-examples.
1	Reason with shapes and attributes Create 2-dimensional shapes or 3-dimensional shapes. Combine shapes to make new shapes. Partition circles and rectangles into two equal shares. (halves, fourths, and quarters)	Ex. Rectangles are 4-sided figures Build and draw shapes based on attributes The rectangle is green colored 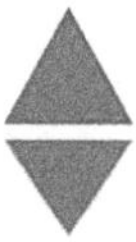  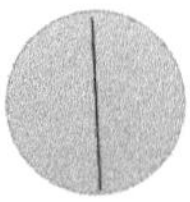 "half"

Figure 12.1. Progression of Geometry Standards Across Elementary Grades
Teruni Lamberg

2	Reason with shapes and their attributes	Draw shapes with specific attributes (Faces, the number of angles) Triangles, quadrilaterals, pentagons, hexagons, and cubes
	Partition a Rectangle and Circle	Split a rectangle into equal squares and count the squares (helping students develop prior knowledge for finding area).
	Partition circles and rectangles	Split into two or three equal shares (halves, thirds, and described as two halves, three thirds, and four fourths.)
3	Shapes and their attribute	*The right angle* is 90 degrees and made up of perpendicular lines 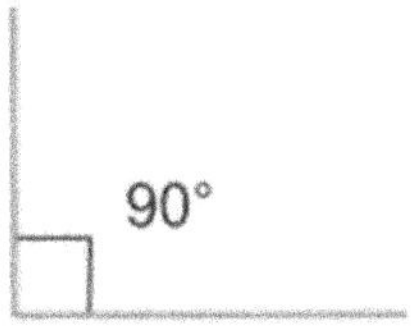 *Parallel lines* do not touch  *Quadrilaterals – 4-sided shapes*
	Quadrilaterals	*Quadrilaterals – 4-sided shapes* Such as quadrilateral, rhombus, square, and rectangle. 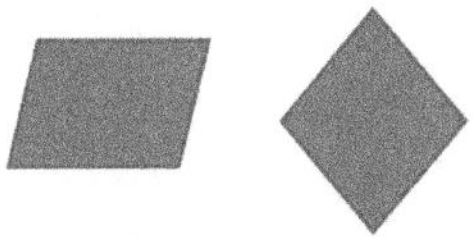

Figure 12.1. Progression of Geometry Standards Across Elementary Grades
Teruni Lamberg

4	Identify angles and lines	Draw points, lines, line segments, rays, and angles (Right acute obtuse) Perpendicular and parallel lines.
5	Graph points on a coordinate plane to solve real-world mathematical problems Classify two-dimensional figures based on their properties	

Figure 12.1. Progression of Geometry Standards Across Elementary Grades
Teruni Lamberg

SHAPES AND THEIR ATTRIBUTES

Parallelograms

- 4-sided shape. These shapes have two pairs of parallel sides, and the opposite angles are equal.

Rectangles
- Have four right angles and two parallel sides (opposite sides are equal.)

Squares: Has four sides with the same length, four right angles, and two pairs of parallel sides.

Trapezoids: A trapezoid is a quadrilateral that has one pair of parallel sides

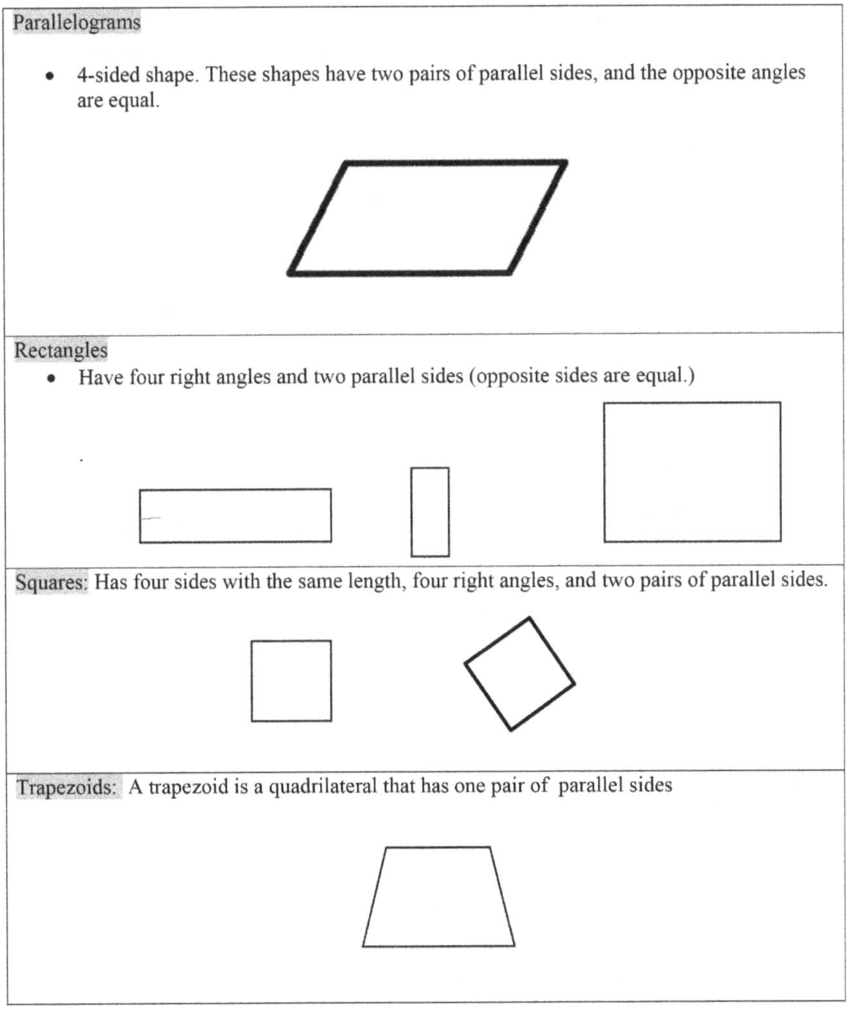

Figure 12.2. Shapes and Their Attributes
Teruni Lamberg

Chapter Twelve

LINES AND ANGLES

Learning about Angles	Example
Point	•
Line Segment	———•———
Ray	 ————————▶
Angle	Right, 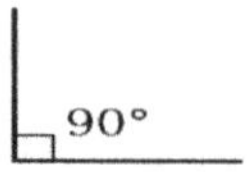 └ 90° acute  obtuse (greater than 90 degrees) Perpendicular (intersect at the right angle) 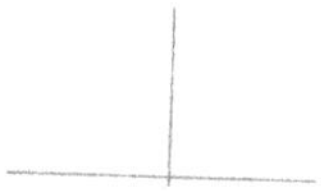

Figure 12.3. Lines and Angles
Teruni Lamberg

BIBLIOGRAPHY

Achieve the Core. (2018). Focus by grade level (achieve the core). https:// achievethecore.org/category/774/mathematics-focus-by-grade-level

Achieve the Core. (2018). Mathematical routines. https://achievethecore.org/content/ upload/Mathematical%20Routines.pdf

California Department of Education. (2015). *Mathematics framework for California public schools: Kindergarten through grade twelve.* www.cde.ca.gov/ci/ma/cf/ documents/transitionalkinder.pdf

Common Core Standards Writing Team. (2019). *Progressions for the Common Core state standards for mathematics* Institute for Mathematics and Education. http:// ime.math.arizona.edu/progressions

Daro, P., Zimba, J., et al. (2019). *Progression documents for the Common Core math standards.* The Institute for Mathematics and Education at the University of Arizona. www.math.arizona.edu/~ime/progressions/

Gravemeijer, K. 1997. Mediating between concrete and abstract. In T. Nunes and P. Bryant (Eds.), *Learning and teaching mathematics: An international perspective* (pp. 315–345). Psychology Press Ltd.

Math Playground. Give your brain a workout. www.mathplayground.com/.

National Governors Association Center for Best Practices, Council of Chief State School Officers. (2010). *Common Core state standards: Math.*

NOTES

Appendix A

Interview Protocol Used with High-Achieving Individuals in This Study

1. What are some things your parents did to support you in learning math?
2. How did they structure study time?
3. What environment did your parents create to help you succeed?
4. What are some of the challenges you encountered learning math in school, and how did you overcome these obstacles? What did your parents do to support you?
5. What resources did you use at home, and how did you get support from your parents?
6. What made you successful in learning STEM subjects?
7. What advice would you give to parents trying to help their children succeed in math in school?
8. Provide any other advice you may have or significant events in your life that shaped your academic journey.

Appendix B

Interview Protocol Used with Parents of High-Achieving Students in This Study

1. What are some things you did as a parent to support your child to learn math?
2. How did you structure study time?
3. What kinds of interactions with your child were helpful them becoming successful?
4. What are some of the challenges you encountered in supporting your child?
5. What resources did you use, and how did you get support for your child?
6. What kinds of interactions did you have with teachers and schools?
7. What advice would you give to other parents?
8. How did you balance the demands of parenthood and your schedule?

Index

ability: grades relation to, 15, 35,
38–39; hyper-focus, 29; instruction
built on, 25

abstract representation, quantitative
representation compared to, 58

addition, *76*, *84*; data analysis in, 99; of
fractions, 93; properties of, *90*; with
regrouping, 5, *5*

ADHD, hyper-focus ability of, 29

algebraic thinking, patterns in, 83, 90

algorithm: borrowing and carrying, 4, *4*;
formal rules as, 75

angles, lines and, *112*

anxiety: from tests, 67; time relation to,
23–24; from visual discomfort, 49

area model, 96, *96*

assessment, formative, 21–22

assessment data: teachers collecting, 38;
understanding of, 42

Autonomous Learner Model, in gifted
program, 13

back-to-school night, goals for, 31–32

bar graph, *104*

beauty: in logical thinking, 7–9, 10; of
problem solving, 49

Betts, George, 13

big ideas, in problem solving, 61

big picture, in problem solving, 15–16

body, mind relation to, 52–53

borrowing and carrying algorithm,
4, *4*

brain: learning effect on, 65, 66; math,
20, *26*; practice effect on, 69

calculus, epsilon delta in, 7–8

capabilities, of children, 28–29

Carraher, T. N., 6–7

categorical data, 103

challenge level, for success, 16–17

challenges, thinking for, 17

chess, mathematics compared to, 3–4

children: capabilities of, 28–29;
motivation of, 45; parents' relation
to, 30–31, 35, 38, 47–48, 63;
perspective of, 21

circle, area of, cylinder relation to,
66–67

classrooms: communication in, 58;
environment of, 33–34

collaboration: environments for, 60, 62;
instead of competition, 57–58,
61–62

Common Core Standards for
Mathematical Practice, xii, 14, 23,
58–59, 60; communication in, 37, 62;
"new math" in, xi; parent knowledge
of, 37; progression in, 75

Nathan, Mitchell, 53
National Library of Virtual
  Manipulatives (website), 80
National Research Council, 4; on
  learning, 66–67; on patterns, 6
natural setting, school compared to,
  6–7
needs, of students, 29, *40*
negative experiences, effect on
  confidence, 19–21
"new math," in Common Core
  Mathematics Standards, xi
notation, standard math, 5
novices, experts compared to, 6
Numberblocks (YouTube videos), 80
numbers: lines of, 96–97; and
  operations, *86–89*; relationships
  between, 83
numerical data, 103

operations, and numbers, *86–89*

parents, *42*; children relation to, 30–31,
  35, 38, 47–48, 63; Common Core
  Mathematics Standards knowledge
  of, 37; encouragement from, 14;
  fixed mindset of, 22; goals for,
  10; homework help by, xi, 38, 46;
  insight of, 28–29; student relation to,
  68; study skill ideas for, *54*; teacher
  relationship with, xii, xiii, 27–28,
  31–33, 35, 39; as volunteers, 33–34
parent-teacher conferences:
  communication in, 27–29, 32, *36*;
  expectations in, 31, 33
patterns: in algebraic thinking, 83,
  90; in data analysis, 99; gifted
  students recognizing, 9; in problem
  solving, 7; recognition of, 67, *68*; for
  understanding, 6, 60
perspective, of children, 21
physical activity, for focus, 52–53
picture graph, *103*
pivotal events, in learning math, 20–21
plan of action, for problem solving, 18

practice: for learning, 52, 69–70; of
  multiplication, 77
precision, in communication, 60
prescriptive approach, of teachers, 20
pride, in final products, 49
problem solving: beauty of, 49; big
  ideas in, 61; big picture in, 15–16;
  in Common Core Standards for
  Mathematical Practice, 58; critical
  thinking skills in, 69; feelings of,
  3–4; joy in, 7; learning for, 68;
  strategies for, *25*; structure in, 8;
  textbooks for, 52; time for, 24; tools
  for, 59; understanding in, 17–18
progression: of fractions, *94–95*; across
  grade levels, 75, *76*; logical, 10; of
  multiplication, *79*, *80*
puzzles, mathematics compared to, 3–4,
  7, 10, 22

quantitative representation, abstract
  representation compared to, 58

rectangle, area of, cylinder relation to,
  66–67
regrouping, addition with, 5, *5*
relationships: as goal, 29; mathematical,
  8, 9; between numbers, 83; parent-
  teacher, xii, xiii, 27–28, 31–33, 35,
  39
respect: mutual, 35; for teachers, 34
results, expectations relation to, 14
rough drafts, of math problems, 49, *50*
routines: to build community, 60, 62;
  for homework, 47–48

school, natural setting compared to, 6–7
schoolwork, games compared to, 6–7
science, technology, engineering, and
  math (STEM), math for, xi–xii, xiii
self-diagnosis, confidence for, 16
self-esteem, negative experience effect
  on, 19–21
sense-making, for discussion, 61, 62
set model, 96, *96*

shapes, *111*; in elementary grades, 105
social interaction, effect on learning,
  57–58
Sparking the Math Brain (Lamberg), xii
spatial reasoning abilities, of gifted
  students, 9–10
state proficiency, in math, xii
STEM. *See* science, technology,
  engineering, and math
strategies: for communication, xii; for
  learning, xiii–xiv, 51; for motivation,
  25; for multiplication, 77, *78*;
  problem solving, *25*
stress: time relation to, 47, 77; from
  visual discomfort, 49
structure, in problem solving, 8
students, 68; expectations for, 48;
  feedback effect on, 40–41; fixed
  mindset of, 21–22; gifted, 9–10;
  goals of, 47; interests of, 13–14;
  needs of, 29, *40*; time of, 42. *See
  also* children
study schedule, 47
study skills, *54*
study space, 47
subtraction, *76*, *77*, *84*; data analysis in,
  99; of fractions, 93
success: fulfilled life as, xii; right
  challenge level for, 16–17
support: of children, 30–31; for learning
  disabilities, 29

teachers, *42*, 68; assessment data
  collected by, 38; feedback of, 40–41;
  formative assessment of, 21–22;
  goals for, *10*; parent relationship
  with, xii, xiii, 27–28, 31–33, 35,
  39; prescriptive approach of, 20;
  respect for, 34; self-directed learning
  facilitated by, 13–14; study skill
  ideas for, *54*
technical job, math for, xi–xii

test-taking skills, understanding
  compared to, 67
textbooks: digital resources for, 38; in
  problem solving, 52; respect for, 34
thinking: algebraic, 83, 90; for
  challenges, 17; emotions effect on,
  70; logical, 7–9, 10; in problem
  solving, 68; time for, 24
third grade: fractions in, 93; geometry
  in, 107
time: anxiety relation to, 23–24;
  homework effect on, 46; stress
  relation to, 47, 77; of students, 42
tools, for problem solving, 59
transfer, learning relation to, 6
trust, between parents and teachers, 33

understanding: of assessment data, 42;
  big picture for, 15–16; formative
  assessment for, 21–22; knowledge
  relation to, 69; learning relation to, 4,
  5, 10, 66–67; memorization without,
  6, 14; patterns for, 6, 60; in problem
  solving, 17–18

variables, in algebraic thinking, 83
visual discomfort, anxiety from, 49
visualization, *19*; in real-world context,
  18
volunteers, parents as, 33–34
Vygotsky, Lev, 23

websites, for math learning, 80
Work Smarter, Not Harder (Lamberg),
  23, 60

Yale medical school, 57
YouTube videos, for math learning, 80

Zone of Proximal Development,
  frustration compared to, 23

About the Author

Teruni D. Lamberg is an associate professor of elementary education at the University of Nevada, Reno. She teaches graduate and undergraduate mathematics education courses. She is the director of the Nevada Mathematics Initiative and the principal investigator and director of the Lemelson Math and Science Master's Cohort Program. She also served as the principal investigator of the Northeastern Nevada Mathematics Project. Lamberg has worked with hundreds of teachers across the country to improve math teaching. Her mission is to provide teachers with tools and resources to support student learning through whole-class discussions.

She spent many years in classrooms researching how to support teachers to conduct effective discussions. She discovered the importance of attending to the process of teaching to facilitate productive discussions that result in student achievement. Lamberg taught elementary school before receiving her doctorate from Arizona State University and completing her postdoctorate work at Vanderbilt University. She served as chair and co-chair of the Psychology of Mathematics Education Northern American Chapter. In addition, she was elected to serve as junior chair/chair of the American Educational Research Association Sig Research in Mathematics Education.

CPSIA information can be obtained
at www.ICGtesting.com
Printed in the USA
BVHW040403140223
657678BV00002B/7